HUMAN

RELATIONSHIPS

an introduction
to sociological concepts

Myron H. Levenson
Indiana University of Pennsylvania

Prentice-Hall, Inc. Englewood Cliffs, New Jersey

Library of Congress Cataloging in Publication Data

LEVENSON, MYRON H.
    Human relationships; an introduction to sociological
concepts.

    Includes bibliographies.
    SUMMARY: An introduction to the basic concepts
of sociology such as the challenge, status, functions, rules, and dynamics of human relationships.
    1. Sociology. [1. Sociology] I. Title.
H95.L45   301   73–13758
ISBN   0–13–446450–8

To
M. C. ELMER, W. E. EDWARDS, H. L. SMITH,
and *especially*
BEVERLY L. LEVENSON

In memoriam to E.S.L. and P.R.S.

Printed in the United States of America
10 9 8 7 6 5 4 3 2 1

Prentice-Hall International, Inc., London
Prentice-Hall of Australia, Pty. Ltd., Sydney
Prentice-Hall of Canada, Ltd., Toronto
Prentice-Hall of India Private Limited, New Delhi
Prentice-Hall of Japan, Inc., Tokyo

CONTENTS

FOREWORD   ix

1

THE FIELD OF INQUIRY   1

Why Study Sociology? *2*
Relationship of Sociology With Other Social
    Sciences, *3*
Sociological Approaches, *4*
Problem Questions, *8*
Selected Adjunct Readings, *8*

2

THE FOCI OF INQUIRY: SOCIAL INTERACTION AND SOCIAL SYSTEMS 11

Social Interaction, *12*
Social Systems, *14*
Vital Data for Analysis, *17*
Problem Questions, *22*
Selected Adjunct Readings, *23*

3

MODES OF INTERACTION: SOCIAL FUNCTIONS AND SOCIAL PROCESSES 26

Social Functions, *27*
Social Processes, *30*
Problem Questions, *34*
Selected Adjunct Readings, *34*

4

LEARNING AND ADJUSTING: CULTURE, SOCIALIZATION, AND PERSONALITY 37

Requirements for Human Life: Culture, *38*
Environment and Culture, *38*
Learning to be Human: Socialization, *40*
Acting as a Social Unit: Personality, *43*
Problem Questions, *45*
Selected Adjunct Readings, *46*

5

THE STANDARDIZATION OF BEHAVIOR: VALUES, ATTITUDES, NORMS AND SANCTIONS 48

Social Control: Norms and Sanctions, 52
Problem Questions, 54
Selected Adjunct Readings, 54

6

THE COMPOSITION OF SOCIAL SYSTEMS: SOCIAL STATUS AND SOCIAL STRUCTURE 57

Social Status, 58
Social Structure, 61
Functional and Invidious Stratification, 66
Problem Questions, 69
Selected Adjunct Readings, 70

7

THE ORGANIZATION OF HUMAN ACTIVITY: SOCIAL ROLES AND SOCIAL INSTITUTIONS 72

Social Roles, 73
Social Organization and Institutions, 75
Institutionalization, Bureaucracy, and Bureaucritization, 77
Problem Questions, 80
Selected Adjunct Readings, 80

8

THE BACKGROUND OF SOCIAL CHALLENGE: SOCIAL DEVIANCY AND SOCIAL DISORGANIZATION   83

Social Deviancy, *84*
Social Disorganization, *87*
Problem Questions, *89*
Selected Adjunct Readings, *89*

9

PATTERNS OF SOCIAL DYNAMICS: SOCIAL CHANGE AND SOCIAL TRENDS   92

Social Change, *93*
Some Theories About Social Change, *99*
Social Trends, *104*
Problem Questions, *105*
Selected Adjunct Readings, *106*

APPENDIX   108

INDEX   115

FOREWORD

This book has been written with three points in mind. First, to present the student, most often a bewildered freshman or a struggling sophomore, with a compact and relatively inexpensive basic text.

Second, I hope to give the student a reasonable introduction to modern sociological thinking rather than a hodgepodge of sociological concepts out of context, confusing charts and diagrams, all too brief basic descriptions of substantive areas, or inadequate examples of methodology.

Third, to give the instructor a text having an organization that has developed over a decade of teaching experience but so set up that additional readings can easily be adopted according to the instructor's preference.

If the author has succeeded in his task it is because he has had great mentors.* If he has not, he alone bears the blame, but perhaps one of his students will do better.

---

*Among them, in alphabetical order, are: W. E. Edwards, M. C. Elmer, J. P. Gillin, J. J. Honigmann, G. B. Johnson, R. L. Simpson, H. L. Smith, R. B. Vance, and V. C. Wright.

1

THE FIELD OF INQUIRY

Are today's young people very different from their fore-bears? Look what were some topics of concern in 1935!

... In recent years there has come to be widespread questioning of the right of states to conscript citizens for military service; university students challenge the R.O.T.C. . . . . Heated controversy is heard everywhere over the merits of birth control.*

*Stuart A. Queen, Walter B. Bodenhafer, and Ernest B. Harper, *Social Organization and Disorganization* (New York: Thomas Y. Crowell Company, 1935), p. 35. Copyright 1935 by Thomas Y. Crowell Co. and printed by permission of the publisher.

Why Study Sociology?

SOCIOLOGISTS ARE CONCERNED WITH EVERYTHING that affects human life and the relationships which exist between and among all human beings. In this present era of human history there is so much that is exciting, disturbing, and even threatening in human society—war, racial confrontation, student unrest, overpopulation, environmental pollution, inadequate housing, etc.—that we could say that sociology might be the most relevant of all of man's pursuits of knowledge. It is, after all, a central concern of sociologists to scientifically determine and analyze the underlying factors in various kinds of human interrelationships. These interrelationships rather than the individuals who participate in them are the main focus of attention. Sociologists view individuals primarily in terms of their memberships in all kinds of interacting groups. These interacting groups, or social systems as we shall call them, have important influence on human behavior and, hence, are necessary objects of study for the understanding of social interaction. We can define sociology, then, as *the scientific study of the social interaction of human beings viewed, not as individuals, but as members of social systems.* This definition implies that sociology has no necessary connection with problems of human society such as those mentioned above—and that is true! Since sociology is concerned with all aspects of human interaction it pays attention to the normal, everyday conditions of life and their associated routine behaviors as well as to the peculiar, disturbing, and abnormal conditions which produce unusual or exaggerated forms of behavior. We have to distinguish between *social problems* (such as delinquency, addiction, widespread poverty, or political extremism) and *sociological problems* which include the former but also deal with mundane matters (such as household chores, courtship practices, or leisure activities) because sociologists are interested in all forms of interaction.

Sociologists are very much interested in studying and developing solutions for social problems, but we have learned through much experience that it is necessary to get the facts and eliminate the myths before we can attempt to devise solutions. Years of study and the lessons of history have revealed that the complex problems of human society are not explainable or solvable in terms of any simplistic ideas of cause and effect. This matter of simple explanations and, hence, simple solutions for complex problems has always been a characteristic of extremist political groups and are major tenets of both the far left and the far right today. It is interesting to note that sociology (and any other form of objective study of human behavior) was suppressed in Nazi Germany, and what is called sociology in the Soviet Union consists primarily of attempts to fit social facts within the framework of Marxist ideology. Totalitarian viewpoints and a concern for obtaining objective knowledge of human social behavior are antithetical to one another.

However, through much arduous effort sociologists have developed a number of useful concepts which help them to comprehend many aspects of human social life. It is an intent of this book to provide the reader with a basic knowledge of a number of those concepts.

Relationship of Sociology With Other Social Sciences

Sociology is a general social science in that it concentrates on patterns of human interaction within and between any kind of social system, under any and all conditions. Some of the other social sciences, though, emphasize the study of human behavior within special contexts or they tend to deal with all behavior even including that of noninteracting individuals. For example, Economics is concerned with man's behavior in relation to making a living by acquiring and distributing needed goods and services; while Political Science is concerned with the organizing and governing of his societies. Psychology, with a frame of reference based on a concern with behavior in its most general sense, is as much interested in individuals as in groups and in man as a biological organism as well as a social entity. Anthropology, which is sometimes referred to as "the science of man," does indeed cover a lot of ground. Anthropologists pay attention to all of man's lifeways, the development of human culture, and the evolution of man as creature, as well as his societies and his material and nonmaterial creations. Geography concentrates on the matter of man's distribution on the earth and his relationship to the varying forms of environment he encounters. Sociology is connected in some way with all the other social sciences. It shares concerns with economics on such matters as problems of production and consumption especially when these relate to such conditions as our present "poverty amid affluence" in the United States. Sociologists are as much interested in social control as political scientists and particularly in these days when questions of "excessive force" or "credibility gaps" occur. Sociology shares with psychology interests in such things as the study of personality development, and the behavior of mobs. It is linked with anthropology in the evaluation of the effect of cultural differences on behavior and it is equally involved with geography in the study of human ecology. All of the social sciences overlap at certain points with the others; however, various theoretical emphases and methods of acquiring and analyzing data tend to distinguish one field from another as well as do their substantive specializations.

Sociological Approaches

There are two general viewpoints that sociologists use in their work. These are not mutually exclusive although some sociologists may strongly emphasize one of them in preference to the other. These two approaches are (1) the basic science approach and (2) the applicative approach. The first of these is primarily concerned with determining the facts of human social behavior. Scientific techniques of collecting and analyzing data are at the core of this approach and a sophisticated methodology has consequently been developed in social research. The use of recording devices, computers, and some highly complex procedures in mathematical statistics are common. The sociologist who uses the basic science approach wants to find out "what is?," that is, the "facts" of human interaction so that he can adequately understand and, perhaps, explain the various kinds of social relationships that human beings maintain with one another.

The sociologist who uses the applicative approach is mainly interested in improving the society of which he is a member. He wants to change the quality of human life in some way which he considers to be better than ongoing conditions represent. He may accept or reject the findings of his more scientifically minded colleagues depending on the kind of orientation he has towards his society. In short, political ideology plays an important part in the approach of the applicative sociologist and it matters not whether he leans toward the right or the left of the political spectrum. Sociologists who use the applicative approach generally attempt to find ways of coping with human problems which they feel are just and reasonable, consonant with their political views, and have no necessary interest in deriving basic facts regarding the general nature of human behavior.

Many sociologists combine both of these approaches to some extent in their work quite often by examining some social problem scientifically with the intent of obtaining findings which can serve two purposes: (1) to provide basic information about the problem and (2) to guide them in providing suggestions for coping with the problem or, most hopefully, to help eliminate it altogether.

We will now have a look at three examples of sociological work. One which represents the basic science approach, another representing the applicative approach, and the third presenting a combination of the first two.

The first of our examples deals with political behavior. The author primarily had the intent of comparing two theories of political "cross-pressure."[1] Patrick Horan who conducted the research was basically interested in determining whether or not patterns of non-voting could be explained by either theory. One of those theories holds that political behavior, such as nonvoting, is explainable in terms of the effects of differences in the social positions (for example, educational level, religious preference, occupation, etc.) held by the individual voters. For example, the implication here is that, perhaps, the condition of being poorly educated, of Roman Catholic faith, and in a low income occupational group, would represent a set of social categories each of which would have some distinctive effect on a person's potentiality to be willing to go to the polls or not. The other theory is that the potentiality for nonvoting, as a type of political

---

[1]Patrick M. Horan, "Social Positions and Political Cross-Pressures: A Re-examination," *American Sociological Review*, XXXVI (August, 1971), 650–60.

behavior, is related to inconsistencies in the social positions held by voters. For example, if certain voters were found to be both highly educated yet only in low income occupations the resultant effect could be a conflict of orientation between their intellectual and their economic sentiments thus leading to a possible desire to escape from such a confusing situation by simply not going to the polls on election day. While both of these theories may have some validity under some circumstances Horan found neither one to be very satisfactory as general explanations for a lack of voting under most circumstances. He made no attempt, however, to provide an alternative theory or any suggestions for a solution to the problem of voters who don't go to the polls.[2] His research was essentially that of the basic science approach, an attempt to gain knowledge without any necessary concern for applying it to a given problem.

Harry C. Bredemeier provides an example of sociological work which stands in contrast to that by Horan.[3] In this article Bredemeier devised a plan for making social welfare programs more efficient. Basically what Bredemeier suggests is that we move to the "strategy of increasing demand"[4] by which he means that urgent public needs be met by having the government supply contracts to private companies for the improvement of housing, recreation, and other public services. These companies in turn would hire many people who are now unemployed including some of those able to work who are presently on relief rolls. In this way the "army of the poor"[5] would be involved in improving their own lives by acquiring jobs, increasing their income, and generally raising their standards of living. In turn this would mean more people contributing to the general economy and thus a rise in the total level of our national prosperity. Bredemeier, in this piece of work, shows no evidence of research conducted; he presented no questionnaires, conducted no interviews, and acquired no data subject to mathematical analysis.

[2]*Ibid.*, 658–59.
[3]Harry C. Bredemeier, "New Strategies for the War on Poverty: A Plan to Enlist Private Enterprise in the Management of Social Welfare," *Trans-action*, (November/December, 1964), pp. 3–8, as reprinted in Harry Gold and Frank R. Scarpitti (eds.), *Combatting Social Problems: Techniques of Intervention* (New York: Holt, Rinehart, Winston, Inc., 1967), pp. 43–50.
[4]*Ibid.*, Gold and Scarpitti, pp. 45–47.
[5]*Ibid.*, pp. 47–49.

He simply looked over the major facets of one of the big problems of our society and, in part, by means of his knowledge of social behavior, and the use of a certain pattern of logic, seemingly based on American style commercial pragmatism, he makes suggestions which he feels will materially aid in resolving the poverty situation more efficiently than has been the case in the past. This is a purely applicative approach for a sociologist to utilize and in no way is there any implication of an attempt to acquire basic facts for a better comprehension of the nature of human society.

An example of sociological literature which presents a combination of both the basic science approach and the applicative approach is to be found in a piece by H. Lever who reports on a social-psychological experiment which was conducted in the Union of South Africa.[6] By using procedures which involved setting up three experimental groups (which are those that receive stimuli whose effects are to be tested) and one control group (which does not receive the forementioned stimuli) Lever was able to demonstrate that it is possible to reduce negative interracial attitudes in a comparatively short period of time. As his subjects Lever used 320 South African college students (mostly whites) and those who were in the experimental groups were all presented with half-hour lectures by speakers who dealt with the subject of race and intelligence. In contrast, the subjects who made up the control group heard only a half-hour lecture on the history of sociology. Scores for social contact (the degree of willingness to associate with others—generally those unlike oneself) were found to be noticeably different between the experimental groups and the control group. In all cases the experimental groups had scores which revealed more desire for contact with others than did the experimental group. Lever thus used a basic science approach to acquire information and to attempt to reduce intergroup tension. He successfully combined both the basic science and the applicative approaches in his work.

These three examples of different viewpoints taken by sociologists in their work reveal that (1) sociologists may, and often do, differ in the way they deal with human groupings; and (2)

---

[6]H. Lever, "Reducing Social Distances in South Africa," *Sociology and Social Research*, LI (July, 1967), 494–502.

that sociologists are capable of combining different approaches without difficulty or confusion. This, of course, is not a phenomenon peculiar to sociology or any other branch of the social or natural sciences but is presented here as an exemplification of the fact that one or another or a combination of viewpoints is neither right nor wrong but rather is dependent upon the nature of the subject matter to be dealt with and the purpose a sociologist has in mind at the time.

Problem Questions

1. Select an introductory text on anthropology and another on psychology from your library. Examine their tables of contents. In what important respects do they differ? In what ways are they similar? Make the same comparison with your sociology text. Is there anything which indicates a linkage among the three fields which you can determine from the tables of contents? What might be a better way to determine the interrelationships of the three disciplines without having to read all the books completely through? (Hint: what is the emphasis of this book?)

2. Poverty is a major area of concern today. Search the periodical indexes in your library for half a dozen articles on the analysis of poverty which have been published in the past decade. Note the major points made by the authors of each article. Are there any important areas of agreement or disagreement? If the articles are selected from nonprofessional periodicals, do the writers mention any professional sources for their opinions or their data? If the articles were obtained from professional journals, are there any noticeable differences in the kinds of data utilized?

Selected Adjunct Readings

PART I—ANNOTATED

Coser, Lewis A. "Social Involvement or Scientific Detachment— The Sociologist's Dilemma," *Antioch Review*, XXVIII (Spring, 1968), 108–13.

In regard to important issues of the day Prof. Coser points out that sociologists have a difficult (and sometimes painful) choice to make with respect to their personal stance as professionals. The basic implication of this article is that there is not a real dichotomy between involvement and objectivity but rather a situation exists wherein the individual sociologist should recognize both his public and his professional commitment. Prof. Coser suggests that it is possible to combine the two orientations without necessarily contaminating either one.

HAUSKNECHT, MURRAY. "Values and Mainstream Sociology: Some Functions of Ideology for Theory," *American Behavioral Scientist,* IX (February, 1966), 30–32.

Prof. Hausknecht illustrates the utility of personal ideological commitments for the development of theoretical constructs in sociology. He does not imply that only an ideological approach to social analysis should be taken but rather that ideology should be complementary to objectivism. He says that any ideological approach can be tested by the operation of scientific logic and procedure which can help to determine its relevancy for explanation of social phenomena.

SCHRAG, CLARENCE. "Philosophical Issues in the Science of Sociology," *Sociology and Social Research,* LI (April, 1967), 361–72.

Prof. Schrag presents an analysis of theory development in sociology that is concise but not obtuse. He describes four criteria of assessment for theories: logical adequacy, operation adequacy, empirical adequacy, and pragmatism. The last is concerned with the relationship of a theory to modes of coping with problems of social policy and values. This article is *must* reading for students in any of the social sciences.

WEINSTEIN, MICHAEL A. "Hocking's Existential Sociology," *Sociology and Social Research,* III (July, 1968), 406–15.

Weinstein presents a valuable description of William E. Hocking's ideas for an existential style of sociology. He notes that Hocking dislikes the concept of sociology as science in terms of the application of scientific method to data he believes not susceptible to such treatment; and because of scientists' (presumed) disregard for moral issues because of their emphasis on the discovery of laws instead of means for action. Essentially, Hocking is presented as a man trying to study American life on the basis of its emotional aspects. To Hocking despair is a key concept for the study of the formation of modern social groups.

WEISS, ROBERT S. "Alternative Approaches to the Study of Complex Situations,"*Human Organization,* XXV (Fall, 1966), 198–206.

Two basic approaches in social science are discussed—the analytic and the holistic. The former takes the standpoint of the isolation of elements and the latter takes the tack of system relationships. Prof. Weiss notes that the two approaches are not completely incompatible and that it is possible to combine both viewpoints into one useful model. He concludes, however, that it is not always possible to translate the results from one approach into terms of the other. This is a good article for students interested in theory construction.

PART II—ADDITIONAL READINGS

BIERSTEDT, ROBERT. "Sociology as Humane Learning," *American Sociological Review*, XXV (February, 1960), 3–9.

ERIKSON, KAI T. "A Comment on Disguised Observation in Sociology," *Social Problems*, XIV (Winter, 1967), 366–73.

HOMANS, GEORGE C. "Bringing Man Back In," *American Sociological Review*, XXIX (December, 1964), 809–18.

HOROWITZ, I. "Discussion: Professionalism and Disciplinarianism: Two Styles of Sociological Performance," *Philosophy of Science*, XXXI (July, 1964), 275–81.

LABOVITZ, SANFORD. "Some Observations on Measurement and Statistics," *Social Forces*, XXXVI (December, 1967), 151–60.

LAZARSFELD, PAUL F. "The Sociology of Empirical Social Research," *American Sociological Review*, XXVII (December, 1962), 757–67.

LEAR, JOHN. "Do We Need New Rules for Experiments on People," *Saturday Review*, XLIX (February 5, 1966), 61–70.

TAYLOR, STANLEY. "Social Factors and the Validation of Thought," *Social Forces*, XLI (October, 1962), 76–82.

2

THE FOCI OF INQUIRY
social interaction
and social systems

When the term "social" is used in polite discourse it often has only the meaning of the opposite of "individual." The following is a better example of the sociologist's idea of what the term means.

As everyone knows, a group of strangers brought together in some common activity soon acquires an informal and spontaneous kind of organization. It comes to look upon some members as leaders, divides up duties, adopts unwritten norms of behavior, develops an esprit de corps.*

---

*Muzafer Sherif, "Experiments in Group Conflict." *Scientific American*, CXCV (November, 1965), 55. Copyright © 1956 by Scientific American, Inc. All rights reserved.

IN CHAPTER 1 SOCIOLOGY WAS DEFINED as the study of interrelationships among human beings considered as members of social systems. This definition can be broken down into the two components of: (1) interrelationships among humans[1] or, more simply, social interaction, and (2) social systems.

Social Interaction

Social interaction has two basic features: *contact,* or mutual awareness among persons, and *communication,* the transmission of meanings and feelings between people.[2] Therefore, when people interact they first become aware of one another's presence and then transmit ideas and sentiments among themselves. The possibility and ease of communication, however, is dependent upon the nature of the contact situation and the total environment. *Total environment* refers to the entire structure of the environment: *physical objects and conditions; biological organisms and their distribution; and sociocultural elements such as relational networks, ideologies and attitudes, and activity patterns.*[3]

Basic forms of contact potential can be described, in relation to the environment, in terms of kinds of proximity, as follows:[4]

---

[1] There are people who study the social habits of animals but animal behavior is primarily instinctive and animal modes of communication are not capable of dealing with the kinds of ideas humans cope with. An interesting presentation on various kinds of animal "societies" can be found in John T. Bonner, *Cells and Society* (Princeton: Princeton University Press, 1955).

[2] See Joyce O. Hertzler, *Society in Action* (New York: Dryden Press, 1954), pp. 63–66, for a detailed discussion of these concepts.

[3] For a more detailed discussion of total environment see R. M. MacIver, *Society: A Textbook of Sociology* (New York: Rinehart and Company, Inc., 1937), pp. 100–113.

[4] This conceptual scheme is derived from Howard Becker, *Through Values to Social Interpretation* (Durham, N.C.: Duke University Press, 1950), pp. 47–48.

12

(1) *Physical proximity*—having to do with geographical location. For example: the place of residence of one group vis-à-vis that of another in regard to the available routes and distances involved; the kinds of transportation obtainable, and the effect of obstacles, such as mountains or deserts in hindering movement.

(2) *Socio-motivational proximity*—concerned with the matter of personal or group desire to make or continue contact with other persons or groups. For example, inhabitants of a "college town" who are not connected with the college might not desire to have contact with students (or faculty members) because they view them as outsiders, or radicals, or troublemakers, etc.[5]

(3) *Psycho-physiological proximity*—quite simply this is related to the innate and acquired ability of people to perceive and comprehend the human part of their environment. The condition of a person's nervous system and his habitual patterns of reaction are important here.

(4) *Cultural proximity*—this has to do with the effects of group practices, beliefs, and experiences. For example—the matter of language similarities and differences, socially developed variations in conceptions of the nature of the world, and the effects of customs related to intergroup and/or intragroup relations.

These categories of contact potential each represent possible areas of limitation on communication and if communication is prevented from occurring then interaction cannot take place. It is important to understand that we are not concerned here solely with verbal communication. Meanings and feelings may be transmitted by gestures, facial expressions, modes of dress, and in many other ways.[6] It should be kept in mind, however, that some form of communication must exist before we can observe and analyze interaction.

Social interaction is a measurable phenomenon. It is neither ephemeral nor a matter of faith. Hence, interaction among persons is capable of being scientifically investigated since we can both define its properties and develop various ways of comparing

---

[5]Naturally an example like this is sheer fantasy!

[6]A good discussion of communication is presented in Jurgen Ruesch and Gregory Bateson, *Communications: The Social Matrix of Society* (New York: W. W. Norton and Company, 1951).

interaction situations with some mathematical precision. There are three measurable aspects of interaction:[7]

(1) *Frequency*—how often interaction takes place. For example, "we meet for lunch everyday," "our club holds its meetings once a month," "the association conducts its seminars biennially."

(2) *Duration*—the amount of time used in interacting. This can be a single figure which represents a particular instance of interaction or it may require that an average be calculated for a number of occurrences. For example, "they discussed the motion on the floor for half an hour," "our special classes usually last for about three hours."

(3) *Intensity*—the degree or strength of interaction. This is the most difficult aspect of interaction to measure. In some cases we must rely on simply "more or less" scales. For example, "feelings ran very high at the convention hall," or "the discussion was carried on at such a low key that the participants seemed to be uninterested in their topic." In other cases we may be dealing with the number of persons involved and the relative amount of positive or negative feeling held. For example, "a large and very angry crowd gathered outside the courthouse," or "only a few unenthusiastic fans greeted the team as they walked out of the locker room."[8] Intensity may be less precisely measured than the other two factors because of its qualitative nature but it must always be accounted for in any reasonable analysis of interaction.

Social Systems

Social interaction can be limited or casual as in the case of a conversation between two passengers on a public vehicle who may never see one another again or when a person excuses his intrusion as he moves past other patrons in a row of theater seats. Some interaction, however, takes place in a more con-

---

[7]Cf., Eliot D. Chapple and Carleton S. Coon, *Principles of Anthropology* (New York: Henry Holt and Company, 1942), pp.36–42.

[8]For other views of interaction see J. S. Slotkin, *Social Anthropology* (New York: Macmillan Company, 1950), pp. 9–19, and Erwin Goffman, "On Facework: An analysis of Ritual Elements in Social Interaction," *Psychiatry: Journal for the Study of Interpersonal Processes*, XVII (August, 1955), 213–31.

sistent pattern among an arrangement of people who regularly come in contact with one another. In other cases people who rarely meet may have to become rather deeply involved at some time or may have to interact for a fairly lengthy period. For example, members of a family living in the same household and seeing each other day after day exist in a fairly constant state of interaction. On the other hand, some persons who do not often get together, such as members of a jury, may, by force of circumstance, be compelled to interact rather intensively, and for a long period of time. Insofar as we discern situations of *relatively consistent interaction* among particular networks of people we can say that we are dealing with *social systems*. Often social systems involve a development of common patterns of behavior among their members. When such systems consist of members who are associated with given geographical and political boundaries, we usually refer to them as *societies*. Fichter has put it this way:

> From the point of view of the persons who constitute it a society is the largest number of human beings who interact to satisfy their social needs and who share a common culture.[9]

The basic implication of this definition is that *a society is a large scale system* which contains a number of smaller systems. A society may or may not coincide with a nation in terms of geopolitical considerations. The whole population of the United States, for example, may constitute one society or we may break it up into regional components such as "the South," "middle America," etc. Each of these, however, contains simpler or more specifically delimited systems such as neighborhoods, student groups, labor organizations, church congregations, state legislatures, or private clubs. Thus, we have a range of types of social systems which includes the minimum two person group as well as entire societies and which varies also according to frequency, duration, and intensity of interaction. It is also possible to classify social systems in terms of the contexts in which they occur. For example, family systems, work systems, leisure systems, etc.[10]

[9]Joseph H. Fichter, *Sociology* (2nd ed.) (Chicago: University of Chicago Press, 1971), p. 160.
[10]See Joseph P. Monane, *Sociology of Human Systems* (New York: Appleton-Century-Crofts, Inc., 1967) for an excellent discussion of types of systems.

We therefore have three types of system characteristics to consider:

1. *Size*—that is, the number of people involved.
2. *Features of interaction*—that is, frequency, duration, and intensity.
3. *Context*—that is, the kind of activities which serve as the realm of interaction.

When each of these three system characteristics has been defined a specific type of social system has been delineated. Since there is such a considerable range of variation possible in each of the types of characteristics it is obvious that there are many kinds of social systems which can exist. It is possible, however, to simplify matters by using a simple mode of classification. The simplest classification we can use is to take the characteristic of size and list various systems as either small or large. The question we must answer, then, is, where do we draw the line between small systems and large systems? Fortunately some arbitrary lines of demarcation have previously been drawn and we can follow them for our purpose. *Small systems* generally refer to relatively compact and close-knit groups of interacting persons such as families, work teams, and play groups. These groups are usually described as being small (most often less than twenty persons) and having relatively high duration and intensity of interaction—thus bringing in the features of interaction in addition to the factor of numbers.[11] How then do we define the large system? The easiest way, of course, is to take our description of them from our definition of the small systems. That is, a large system consists of a sizeable number of people (at least twenty in contrast to the small systems) but can we say that a large system necessarily involves relatively low duration and intensity of interaction?

With regard to the latter it might be reasonably safe to assume that the more persons who constitute the system units, the less the overall pattern of intensity.[12] However, in the case

[11]See Clovis R. Shepherd, *Small Groups: Some Sociological Perspectives* (San Francisco: Chandler Publishing Company, 1964), pp. 1–6, and cf. Charles Horton Cooley, "Primary Groups and Primary Ideals," in Edgar F. Borgatta and Henry J. Meyer, *Sociological Theory: Present-Day Sociology From the Past* (New York: Alfred A. Knopf, 1956), pp. 226–29.

[12]This essentially is the position of Cooley. See the source noted in footnote 11 above.

of duration the variability in time is too great for us to talk of short duration as a characteristic. Some large systems may be of comparatively short duration. For example, a meeting of cat fanciers may last only one day. But many large systems, such as communities or nation-states, may exist for very long periods of time.[13]

Vital Data For Analysis

In order to comprehend what takes place in a social inter-action situation and to understand social system effects on personal behavior we must first obtain certain vital data.

The Demographic Question

We certainly want to know, first of all, the characteristics of the population which comprises the social system under investigation. This is the *demographic question*. It consists of three parts, the first of which is the matter of the *size and distribution of a population*. Here we are concerned with such things as the total number of persons interacting and the density of a group according to its spatial boundaries. With regard to group size, Georg Simmel long ago noted its effect upon the structure and organization of a social system.[14] For example, he noted that in a two-person group (dyad) each member is absolutely dependent on the other. In a three-person group (triad), however, he suggested that one member could play off the other two against each other in order to achieve his own ends.[15]

Density, on the other hand, has been found to have particular

[13]An interesting discussion of interaction in which rural communities, considered as small scale systems, are contrasted with urban communities viewed as large scale systems is to be found in Pitirim Sorokin and Carle C. Zimmerman, *Principles of Rural-Urban Sociology* (New York: Henry Holt and Company, 1929), pp. 48–58. Also cf. Ferdinand Tonnies, *Gemeinschaft und Gesellschaft*, translated and edited by Charles P. Loomis under the title *Fundamentals of Sociology* (New York: American Book Company, 1940).

[14]See Georg Simmel, "The Number of Members as Determining the Sociological Form of the Group," translated by Albion W. Small, *American Journal of Sociology*, VIII (July, 1902), 1–46 and (September, 1902), 158–96.

[15]*Ibid.*, (September), 163–86.

effects on behavior with noticeable differences between high density and low density groupings.[16] A high degree of occupational specialization, for example, has been found to be more characteristic of densely populated than of sparsely populated areas.[17] Among bands of nomadic hunters where the total population of the area is quite low it has been noted that all the men have much the same tasks, and all the women have similar tasks. They are too few to specialize. Everyone must do a number of things. Furthermore they feel more comfortable in each others presence than among persons of the same origin but from another band.

Calhoun, in laboratory experiments with rats, found that when the rat population was allowed to increase in a confined space, thus raising density, the rats would develop extremely abnormal behavior including cannibalism and homosexuality.[18] Calhoun has also noted the relationship between stress and population growth among humans; his rat study was really a demonstration of the same effect of overcrowding on any kind of mammalian population.[19] It is not unusual to find high crime rates in densely populated areas and lower rates in areas of less density.

The second part of the demographic question consists of the determination of the factors involved in the growth or decline of a population. Here we must examine patterns of fertility, mortality, and migration. That is, we have to look at the changes wrought by births, deaths, and geographic mobility.[20]

Finally, in answering the demographic question, we must look at the composition of the population. Who is interacting with whom? This includes a consideration of such variables as

[16]See, e.g., Nathan Keyfitz, "Population Density and the Style of Social Life," *BioScience*, XVI (December, 1966), 868–73.

[17]*Ibid.*, and cf. Emile Durkheim, *The Division of Labor in Society* (New York: Free Press, paperback edition, 1964), pp. 11–146.

[18]John B. Calhoun, "Population Density and Social Pathology," *Scientific American*, CCVI (February, 1964), 139–48. Also cf. John J. Christian, "Phenomena Associated with Population Density," *Proceedings of the National Academy of Science*, XLVII (April 15, 1961), 428–49.

[19]John B. Calhoun, "Social Welfare as a Variable in Population Dynamics," *Cold Spring Harbor Symposia on Quantitative Biology*, XXII (1957), 339–56.

[20]See, e.g., Angus Campbell et al., *The American Voter* (New York: John Wiley, 1960), especially pp. 231–49 which deal with the effects of migration on political behavior.

age, sex, education, marital status, occupation, income, ethnic background, etc. Certainly a group composed solely of women will have different characteristics of interaction than one composed solely of men. A system made up of a large proportion of young people and very old persons in contrast to the middle age ranges is likely to be behaviorally different than one where persons in the middle age ranges predominate.

Generally speaking, if we take the case of the United States, we can say that persons who are under 25 years of age constitute the young members of the population while persons over 65 make up the very old segment. In 1960, as an example, almost 45 per cent of the American population was under 25 and more than 9 per cent was over 65. The young and the old together thus totalled 54 per cent, or a majority of the population.

The Ecological Question

Another kind of data we must obtain in order to understand systems of interaction are those characteristics of the total environment which affect the interaction patterns being observed. This we can refer to as the *ecological question.* Here we are concerned about five kinds of relationships. The first deals with the relationships between one man or set of men and another, or among several men (or sets of men). In this case the men (or sets of them) are each viewed as representing particular social systems; that is, they serve as units of their systems rather than as individuals. This idea, of course, is the basis of sociology as a distinct discipline. While this thought may be somewhat upsetting to certain people, nevertheless it is the basic consideration which distinguishes sociology from psychology.

Some examples of this relationship are:

1. The *ambassador* from one country presenting his credentials to the *leaders* of another.
2. *Union* and *management* officials discussing terms for settling a strike.
3. A *policeman* telling a *dragster* that he can't race on city streets.

It should be kept in mind that persons representing particular systems may, in fact, also be fellow members of some other, more inclusive, systems. For example, the ambassador and leaders mentioned above would be members of a diplomatic system including the representatives of many other nations.

The second ecological relationship is that between some groups of men and other living things. Here we examine and note the effects of plants, animals, and microorganisms on human beings. In this case, of course, human beings other than those who are the objects of study would have to be accounted for as living organisms.

A couple of examples of this relationship are:

1. The effect of the failure of a food crop on the living standard of a society.
2. The effect of an abundance of ragweed or similar allergy producing weed on the quality of an area as a place of residence.

Yet another example of vast historical importance was the Black Plague which killed millions of people in Medieval Europe. That epidemic involved the transmission of a germ to humans by means of the bites of fleas which lived on the bodies of rats.[21]

The third ecological relationship is that between men and natural physical objects or conditions such as mountains, minerals, the wind, rain and snow, coastal plains, deserts, oceans, volcanoes, etc. We can easily come up with examples such as the discovery of gold in California and the consequent Gold Rush leading to the relatively rapid westward expansion of the United States. We can also look at natural disasters like the eruption of Vesuvius in A.D. 79 which destroyed Pompeii and Herculaneum. Or we can observe the overcoming of obstacles as in the case of the development of navigational instruments by the seafarers and desert nomads of the Middle East. All of the foregoing cases had notable effects on human behavior and led to the modification or disruption of various social systems. However, those were extreme examples. Many of man's adaptations to nature take the form of simple actions, some of which do not become customary behavior for a given group nor have any major effect on the characteristics of a social system. For example, consider the person who covers his eyes with his hands during an electrical storm because he can't stand the sight of lightning. It is doubtful that even in the case of a small system,

---

[21]See Herman Styler, *Plague Fighters* (Philadelphia: Chilton, 1960), or Philip Ziegler, *The Black Death* (New York: J. Day, 1969).

such as that person's family, that every member would follow his way of doing things under such circumstances.

The fourth ecological relationship is that between men and their artifacts; the nature of human interaction is very much affected by man's physical creations such as roads, firearms, highrise apartment houses, microscopes, etc. Sometimes of course, the results are not those which the creators intended—as in the case of Alfred Nobel who hoped that his invention, dynamite, would be used solely for peaceful purposes. Still other creations of man have become so common that they are often notable to people only by their absence—city persons who move to the suburbs sometimes find the lack of curbs and sidewalks to be quite uncomfortable.

Finally, the fifth ecological relationship is the very important one between men and abstract ideas. No doubt many Americans could become very disturbed when someone challenges "free enterprise" or "rugged individualism." Wars have been fought over differences in religious beliefs and it is hard to conceive of the nineteenth century without the "theory of evolution" or "laissez-faire" or "Victorian morals."

It is necessary to understand that all five of the above stated relationships must be considered before the ecological question can be answered.[22] If any are left out there will be gaps in the social analyst's data which may lead him to the errors of misplaced emphases or the drawing of erroneous conclusions about any or all types of human interactions.

The *ecological question* is clearly concerned with man as a communal creature. Since human beings are naturally gregarious (the true hermit is really an exceptional rarity) the various

---

[22]Essentially, these five component items are dealt with in any good ethnography (report of the culture of a given group of people by a field anthropologist). As examples, see Robert T. Anderson and Barbara G. Anderson, *The Vanishing Village: A Danish Maritime Community* (Seattle: University of Washington Press, 1964); Donald D. Brand (assisted by José Corona Nunez), *Quiroga: A Mexican Municipio* (Washington, D.C.: Smithsonian Institution, Institute of Social Anthropology, Publication no. 11, 1951); or, John Gillin, *The Culture of Security in San Carlos* (New Orleans: Middle American Research Institute, Publication no. 16, Tulane University, 1951). For a discussion of the operation of these factors in human interaction see John J. Honigmann, *The World of Man* (New York: Harper and Brothers, 1959), pp. 16–19; 156–68; 345–61; and 439–59. Also, cf. Honigmann's *Understanding Culture* (New York: Harper & Row, 1963), pp. 287–90, wherein he discusses the effects of people and things *in re* socialization.

forms of human collectivities—formal and informal associations, communities, and kin groups—are of great importance as interaction contexts and, thus represent important spheres of attention for sociological investigation.[23]

The Behavioral Question

Last of all, we want to know something about the dynamic nature of the interaction situations we observe. This is the *behavioral question.* To answer it we must attempt to describe and explain:

1. the nature of the activities being conducted.
2. the purposes and consequences of the individual acts being performed.
3. the modes of interpersonal and/or intergroup relations which are going on.

This last question is, perhaps, no more important than the other two but due to its centrality from the conceptual standpoint the next chapter is devoted to much of its core.

Problem Questions

1. Make up a list of at least five social systems of which you are a member. For example, your classes, clubs, fraternity or sorority, athletic teams, etc. Describe each of them in terms of:

[23]See, *e.g.,* James Quinn, *Human Ecology* (New York: Prentice-Hall, 1950); Irwin Sanders, *The Community: An Introduction to a Social System,* 2nd. ed. (New York. Ronald Press, 1966); Robert S. Lynd and Helen M. Lynd, *Middletown* (New York: Harcourt, Brace & World, 1929); W. Lloyd Warner and Paul S. Lunt, *The Social Life of a Modern Community* (New Haven: Yale University Press, 1941); George P. Murdock, *Social Structure* (New York: Macmillan, 1949); Robert H. Lowie, *Social Organization* (New York: Rinehart and Company, 1948); M.G. Smith, *West Indian Family Structure* (Seattle: University of Washington Press, 1962); Mhyra S. Minnis, "Cleavage in Women's Organization: A Reflection of the Social Structure of a City," *American Sociological Review,* XVIII (February, 1953), 47–53. (The preceding is nowhere near inclusive even of major works but represents a sampling of some of the more important ones.)

(a) Frequency, duration, and intensity of interaction.

(b) The *demographic question.*

(c) The *ecological question.*

Having described them can you denote any characteristics they all share? Which two are most unlike one another and in what ways (other than your own membership) are they related?

2. Select two journal articles, each about a different kind of social system, and note what the authors indicate to be the special characteristics of each. Do the authors provide any clues which imply that the characteristics emphasized have anything to do with the number of people involved in each of the systems described? Are the system characteristics in any way dependent upon the age, sex, or race of the members?

Selected Adjunct Readings

PART I—ANNOTATED

GLASER, BARNEY G. and STRAUSS, ANSELM L. "Awareness Contexts and Social Interaction," *American Sociological Review,* XXIX (October, 1964), 669–78.

Professors Glaser and Strauss present a discussion of "awareness contexts" and attempt to provide useful typologies of types of contexts within which interaction occurs. They provide a general definition of an awareness context as [". . . the total combination of what specific people, groups, organizations, communities, or nations know about a specific issue."] Four basic types of these contexts are described: open, closed, suspicion, and pretense. They develop a paradigm which focuses on the interaction process in flux rather than as a static entity and they discuss the contributions of George H. Mead, Erving Goffman, and Fred Davis to the study of interaction. This is a highly interesting, sophisticated, and informative article. Anyone interested in the study of interaction must read it in order to be well informed on interaction theory.

HASSINGER, EDWARD. "Social Relations Between Centralized and Local Social Systems," *Rural Sociology,* XXVI (December, 1961), 354–64.

In looking at the social systems of rural and urban communities Professor Hassinger points up the necessity of examining the relationships between the two kinds of systems as they tend to be merged in our rapidly urbanizing society. He notes the importance of the values of the whole society in mediating those relationships and also pays attention to factors of social structure and community organization. This paper pays special attention to roles and bureaucratic set ups which may provide crucial links between local and larger systems. While necessarily compact, this work contains a great deal of value to the student of community social systems.

HILLERY, GEORGE A., JR. "Villages, Cities, and Total Institutions," *American Sociological Review* XXVIII , 779–91, October, 1963.

Professor Hillery does not accept the vague term "community" as it is currently used by many social scientists. The term is sometimes employed to refer to residential populations and, in a quite different context, is also defined as institutional populations, Hillery prefers the use of two other concepts:

(1) the *vill* refers to every kind of residential grouping ranging from folk villages to cities.

(2) the *total institution* (as defined by Erving Goffman) which covers diverse types of institutionalized collectivities such as prisons or mental hospitals.

Three foci common to the vill are (a) space, (b) family, (c) cooperation.

In the case of the total institution the important characteristics are a bureaucratic organization and the lack of family type groupings.

This article thus distinguishes two major kinds of ecological entities within which particular patterns of interaction are characteristic and between which there is sharp contrast in the usual forms of interaction.

RYDER, NORMAN B. "Notes on the Concept of a Population," *American Journal of Sociology,* LXIX (March, 1964), 447–63.

In this paper the sociologist's view of social systems is seen to be complemented by the demographic approach to social analysis when the dimension of time is emphasized. Professor Ryder views *cohorts* as aggregates of persons who co-occupy a period of time and, hence, represent an analytical unit which can be used for the study of social change.

Ryder implies that proper use of cohort analysis will allow for the determination of both individual and group factors which are important to the changes a society may undergo from one time period to another.

This article should be carefully read by the student who is interested in the effect of population processes on social systems.

PART II—ADDITIONAL READINGS

BALES, ROBERT F. "How People Interact in Conferences," *Scientific American*, CXCII (March, 1955), 31–35.

BORGATTA, EDGAR F. "The Analysis of Patterns of Social Interaction," *Social Forces*, XLIV (September, 1965), 27–34.

COLLINS, ORVIS. "The Elemental Processes of Simple Interaction: A Formulation." *Human Organization*. XXVI (Spring-Summer, 1967), 6–21.

DEAN, DWIGHT G. and BRESNAHAN, B. S. "Ecology, Friendship Patterns, and Divorce: A Research Note," *Journal of Marriage and the Family*, XXXI (August, 1969), 462–63.

ETZIONI, AMITAI. "Man and Society: The Inauthentic Condition," *Human Relations*, XXII (August, 1969), 329–32.

KUNKEL, JOHN H. "Some Behavioral Aspects of the Ecological Approach to Social Organization," *American Journal of Sociology*, LXII (July, 1967), 12–29.

MICKLIN, MICHAEL. "Urban Life and Differential Fertility: Specification of an Aspect of the Theory of Demographic Transition," *Sociological Quarterly*, X (Fall, 1969), 480–500.

PATTERSON, M. "Spatial Factors in Social Interactions," *Human Relations*, XXI (November, 1968), 351–61.

PENALOSA, FERNANDO. "Ecological Organization of the Transitional City: Some Mexican Evidence," *Social Forces*, XLVI (December, 1967), 221–29.

SCHWARTZ, BARRY. "The Social Psychology of Privacy," *American Journal of Sociology*, LXXIII (May, 1968), 741–52.

SELVIN, HANAN C. and HAGSTROM, WARREN O. "The Empirical Classification of Formal Groups," *American Sociological Review*, XXVIII (June, 1963), 399–411.

WATSON, JEANNE and POTTER, ROBERT. "An Analytic Unit for the Study of Interaction," *Human Relations*, XV (August, 1962), 245–63.

WILLHELM, SIDNEY. "The Concept of the Ecological Complex," *American Journal of Economics and Sociology*, XXIII (July, 1964), 241–48.

3

MODES OF INTERACTION

social functions and

social processes

Human social interaction falls into several distinctive patterns, some of which are connected with activities which are necessary to the operation or survival of social systems. Among a particular group of Tibetans, for example:

. . . conciliation was the chief method of settling both civil and criminal disputes in Sakya. Minor civil disputes were usually settled locally and privately with the good offices of some neighbor or friend, or . . . the matter might be brought to the attention of the village headman who would appoint a mediator . . .*

*Dan F. Henderson, "Settlement of Homicide Disputes in Sakya (Tibet)," *American Anthropologist*, LXVI (October, 1964), 1101. Reproduced by permission of the American Anthropological Association.

IN CASES OTHER THAN THOSE where it is necessary to make broad descriptive surveys sociologists more often than not concentrate their research efforts on specific kinds of activities. Some sociologists pay special attention to activities in the health and medical field,[1] others to matters of family relationships,[2] some are concerned about the problems of school desegregation,[3] and so on and on goes the list. It is necessary to keep in mind, however, that regardless of the kind of activities being observed the social analyst needs to account for the *functions* and *processes* which the activities demonstrate.

Social Functions

Social functions refer to acts and attitudes which the members of a social system consider to be necessary for meeting system needs. All systems, regardless of their individual natures, share a number of common social functions which must sometimes be accomplished just for the systems to come into being as well as for maintaining themselves over time. These basic functions which have been called "functional prerequisites"[4]

---

[1]See, e.g., Leo W. Simmons, "Important Sociological Issues and Implications of Scientific Activities in Medicine," *Journal of the American Medical Association*, CLXXIII (May, 1960), 11.

[2]See Alan C. Kerckhoff and Keith E. Davis, "Values Consensus and Need Complementarity in Mate Selection," *American Sociological Review*, XXVII (June, 1962), 295–303, as an example of high-level research in this field.

[3]For good examples in this area see James W. Vander Zanden, " Desegregation and Social Strains in the South," *Journal of Social Issues*, Xᵛ (1959), 53–60 or his "Turmoil in the South," *Journal of Negro Education* (Fall, 1960), pp. 445–452.

[4]See D. F. Aberle et al., "The Functional Prerequisites of a Society," *Ethics*, LX (January, 1950), 100–111.

or functional requisites"[5] include factors such as: recruitment of members, communication channels, meeting of bodily needs, protection from hazards, and maintenance of order among the members.[6] These functions may, respectively, be accomplished by: sexual reproduction—as regulated by marriage customs, or, by adoption; establishment of common signs and symbols—language (oral and written) plus technical aspects of communication (telegraph, telephone, radio, etc.); development of means for obtaining food and water, such as agriculture, well-digging, irrigation or dam building; providing means of shelter and defense such as buildings, clothing, weapons and medicine; and, by originating patterns of leadership, rules of organization and regulations for individual behavior.[7] In addition to the basic functions, however, there are certain specific acts and attitudes which a particular social system may require in order to meet needs that are peculiar to its own situation.

For example, an athletic team may feel obliged to win a certain number of games per season even though its won and lost record may have no effect whatever on its continued existence. The need served in this case may be only that of bolstering the members' *esprit de corps* but for that system at that point in time this need may, in fact, be a very important one. In another case let us suppose we are observing a group of primitive fishermen. If these fishermen have good nets, are highly skilled at casting and retrieving them, and the fish are running in great quantity, we might assume that they need do no more than get to the water and haul in their catch. But they might not see things our way. Instead, they may feel it is necessary to propitiate the water spirits before doing anything else. They might do that by burning last year's nets and dancing around the fire or, as an alternative or perhaps as extra insurance, they might throw one of their village leaders into the water a few times. To the outside observer those practices might appear to

---

[5]See Marion F. Levy, Jr., *The Structure of Society* (Princeton, N.J.: Princeton University Press, 1952), pp. 151–97.

[6]Cf. Joyce O. Hertzler, *Society in Action* (New York: Dryden Press, 1954), pp. 20–22, wherein he discusses the primary functions of a society.

[7]Cf. Gerhard Lenski, *Human Societies: A Macrolevel Introduction to Sociology* (New York: McGraw-Hill Book Company, 1970), pp. 29–30. Also see Don Martindale, *American Society* (Princeton, N.J.: D. Van Nostrand Company, 1960), especially pp. 254–427 where he discusses institutional structures in mass society.

be silly and worthless, but according to the customs of the fishermen the practices could be considered as more important than good nets to insure successful fishing.

Lester F. Ward would have referred to social functions as *telic* (purposeful) behavior—from the group rather than the individual standpoint. *Social telesis* is a concept Ward developed in his first great work, *Dynamic Sociology*, which was published in 1883. Ward recognized, and we are operating here on the assumption, that persons are so conditioned by their societal membership, that beyond some idiosyncrasies, the bulk of their action and thought is oriented to the attainment of things (both concrete and abstract) which their society deems to be desirable or necessary. Their behavior then, is primarily *functional for their social system.* The behavior of any people in this sense does *not* have to be functional for individuals. That is, the behavior, whatever the kind of activity of which it is a part, does not have to meet personal needs in order to satisfy the requirements of the system. For example, if some soldiers are ordered to hold a certain position to prevent an enemy from advancing they may all be killed in the attempt. However, so long as the enemy forces are prevented from advancing for the necessary period of time, then the requirements of the larger military unit (as a social system) will have been met even if the personal survival needs of the individual defending soldiers have not. When any kind of behavior is not considered to be desirable or necessary for a system, then it would be classified as *dysfunctional*—that is, it will be viewed as operating against the satisfaction of the needs and, perhaps, the very existence of the entire system.[8] As an example of a dysfunctional act we could take the case of a ship's navigator who becomes intoxicated and thus is unable to properly plot the ship's course. Such action could lead to disaster for the ship and its crew.

The viewpoint we are taking here of *function*, as a sociological concept,[9] is that it refers to complexes of attitudes and actions which are conducive to accomplishing whatever has been

[8] See Harry M. Johnson, *Sociology: A Systematic Introduction* (New York: Harcourt, Brace & World, Inc., 1960), pp. 63–68, for a very good discussion of function and dysfunction.

[9] Robert K. Merton has discussed different ways of conceptualizing function in his *Social Theory and Social Structure* (Glencoe, Ill.: Free Press, revised edition, 1957), especially pp. 50–55.

defined within a social system as necessary in order to achieve system-defined and approved goals. These goals may be either (1) *short term,* relating to immediate requirements for conducting specific activities—such as the acquisition of a calculator by a team of research scientists; or (2) *long term,* such as the continued pursuit of new knowledge in a field of science—which may be the basic purpose for which a research team has been established.

Social Processes

If we are to adequately comprehend the interaction of members of a given social system we must, minimally, be able to classify their behavior in socially meaningful terms. In other words, we must have some way of defining the interaction context with regard to the nature of the relationships between the parties involved. Around the turn of this century the German sociologist Georg Simmel was writing about the characteristics of groups and ways of defining the behavior of humans as interacting units.[10] To Simmel, *social forms* were the important focus of attention for sociologists. By social forms he meant forms of interaction, such as the relations between persons in terms of domination and subordination, or patterns of conflict.[11] Simmel and a colleague, Leopold von Wiese, both considered the study, classification, and analysis of forms of interaction as the basic representations of human society and thus to be of prime importance.[12] This point was also paid some attention by other sociologists, notably Robert E. Park and Ernest W. Burgess, who discussed *social processes* in their famous *Introduction to Science of Sociology*.[13] Park and Burgess dealt primarily with four social processes; competition, conflict, accommodation, and assimilation.[14] These were defined as follows:

[10]E.g., see Georg Simmel (translated by A. W. Small), "Superiority and Subordination as Subject Matter of Sociology," *American Journal of Sociology,* II (1896–1897), 167–89 and 392–415.

[11]See Kurt H. Wolff (ed.), *The Sociology of George Simmel* (Glencoe Ill.: Free Press, 1950).

[12]See Leopold von Wiese, *Systematic Sociology,* edited and translated by Howard Becker (New York: Wiley, 1922).

[13]Chicago: University of Chicago Press, 1921.

[14]*Ibid.,* pp. 504–784.

(1) *competition*—viewed as the basic process of the economic system, Park and Burgess essentially defined it as a mutual impersonal and continuous struggle to attain a position in the economic order of a community, often *without* social contact.[15]

(2) *conflict*—seen as the basic process of the political system, and defined as a conscious personal struggle for status and control in society, involving direct but intermittent contact among the contestants.[16]

(3) *accommodation*—looked upon as basic to social organization and defined as being the social equivalent of biological adaptation.[17] It has to do with a striving for consensus in a society through the adjustment of sentiments and activities and, sometimes, a change in patterns of organization.[18]

(4) *assimilation*—conceived by Park and Burgess as underlying the linkage between personality and culture.[19] They defined the concept in reference to immigrants to the United States by saying it means the transmission of local culture to an adopted citizen.[20]

Although those concepts have their uses as defined by Park and Burgess they are of insufficient generality for modern sociology. Hence, other and less narrowly restricted definitions have been applied to the terms in more recent years. Notable among the theorists who have dealt with social processes as elementary patterns of interaction is Joyce O. Hertzler. Hertzler has redefined the four processes discussed by Park and Burgess and added two more which altogether make for a *relatively* complete[21] system for classifying types of social interaction. His two additions are *cooperation* and *contravention*.[22] Hertzler refers to cooperation as a basic process of societal organization,

---

[15]*Ibid.*, pp. 504–10 and 554–64. This is the definition used by some students of urban life and was the basic definition of the so-called ecological school, of which Burgess, Park, and others at the University of Chicago were the founders.

[16]*Ibid.*, pp. 574–79 and 641–48.

[17]*Ibid.*, pp. 663–71.

[18]*Ibid.*, pp. 718–28.

[19]*Ibid.*, pp. 734–40 and 769–75.

[20]*Ibid.*, p. 734.

[21]Complete is used here in the sense that all basic types of interaction can be subsumed under the process categories utilized. This, however, may actually be more simple or more complex than is really necessary in some cases.

[22]Hertzler, *Society In Action*, pp. 152–57 and 268–70.

and defines it as ". . . the process by which individuals or groups act jointly . . . in the pursuit of common interests, purposes, and objectives."[23]

Contravention is viewed by Hertzler as an *opposition process* whereby people struggle against one another.[24] He defines the term to mean something midway between competition and conflict; more severe than the former but less severe than the latter.[25]

Let us now take the six processes noted above (competition, conflict, accommodation, assimilation, cooperation, and contravention) and place them in a system of classification which will enable us to use them as primary descriptions of forms of social interactions.[26] To construct this scheme let us first categorize the processes as either *associative*—tending to draw people together (or to have them act in concert) or, as *antagonistic*—tending to cause people to act against one another. Under the *associative* rubric we can place accommodation, cooperation, and assimilation, while competition, contravention, and conflict belong in the *antagonistic* category. Having done this we must now define the processes in appropriate terms. For clarity we will do this by providing both a definition and a reasonable example for each. We thus have the following definitions and examples:

Associative Processes

(1) *Accommodation*—refers to those actions and attitudes by which one or more parties make adjustments in their behavior in order to avoid dissension, or disruption of desired activities and relationships. The adjustments need *not* be made voluntarily.

Example: a young man agreeing to take his date to a particular movie even though he would prefer to see another.

(2) *Cooperation*—refers to mutual action by all parties in order to attain commonly desired ends. Ordinarily this process should be viewed as involving voluntary, that is, noncoerced action.

---

[23]*Ibid.*, p. 152.
[24]*Ibid.*, p. 268.
[25]*Ibid.*, p. 268.
[26]The definitions used here are generally, but not always, paraphrasings of Hertzler's interpretations of the concepts. Consequently, any inadequacies the reader might perceive cannot be attributed to Professor Hertzler.

Example: carpenters, masons, glaziers, etc., working together to build a house.

(3) *Assimilation*—refers to more or less gradual adoption of one party's patterns of behavior by another, or, to the combining of actions and attitudes of a number of parties with the end result that, behaviorally speaking, none of the parties is distinct from the others. Thus, assimilation refers to cultural fusion. This process need not involve voluntary action.

Example: immigrants learning the ways of life of their host society.

Antagonistic Processes

(1) *Competition*—refers to action by two or more parties striving to attain the same or similar goals and often attempting to outdo one another. This process, however, involves limitations on the action taken by each party, hence, requiring some agreement on rules and, thus, involving some degree of cooperation. Strictly speaking then, competition is only partly antagonistic.

Example: athletic games, such as baseball.

(2) *Contravention*—refers to action by one or more parties attempting to prevent others from attaining a goal, *whether or not they desire to attain the same goal*. In short, contravention involves effort to frustrate others. Since direct injury to others is *not* the intention we can say that this process involves limited opposition.

Example: larger child putting a small child's toy out of reach.

(3) *Conflict*—refers to completely unlimited action by any or all parties with the deliberate attempt to do injury to or to wholly eliminate the opponents from the field of action. Conflict involves no restrictions on attitudes or overt behavior whatever. It is the process of total antagonism.

Example: war.

Generally speaking, it may be declared that the six processes defined above can be used, either singly or in combination, to describe any situation of human interaction. It should also be noted that both associative and antagonistic processes may, and often do, occur within any given social situation.

The processes discussed in this chapter represent patterns of social dynamics, that is, they describe the operational aspect of social systems. As such they are important to the study of changes in social systems and should be the object of intensive study so that eventually, given other and additional knowledge, it will be possible to make reasonably accurate predictions about the future of particular systems.

Problem Questions

1. Select a social system of which you are a member, for example: fraternity, athletic team, laboratory section, etc. List the most important activities in which the system members engage. Break the activities down into a set of basic attitudes and required actions. Using one activity as an example, how does it help to achieve system defined goals?

2. Think back to yesterday and pick some social interaction situation in which you took part. Describe what went on in terms of social processes. Did any interaction take place which could not be described by the processes defined in this chapter?

Selected Adjunct Readings

PART I—ANNOTATED

FIELLIN, ALAN, "The Functions of Informal Groups in Legislative Institutions," *The Journal of Politics*, XXIV (February, 1962), 72–91.

Professor Fiellin presents a good illustration of the concept of function in relation to restricted systems in the form of legislatures. He notes that informal groups which arise in unstructured situations tend to act in ways which may either contribute to the operation of a legislative system (and thus are *functional* for the system), or the groups may take actions which, while they may be useful for the members of the group, tend to reduce the operational efficiency of the system (and, hence, are *dysfunctional* for the system).

An example is given of the "Tuesday-Thursday Club" referring to a short work week for some congressmen, thus causing them to acquire an undersirable reputation which may be harmful to their careers. On the other hand, the author shows how membership in some informal groups may help a congressman to exert influence he might not otherwise be able to acquire.

HEISS, JERALD. "Factors Related to Immigrant Assimilation: The Early Post-Migration Situation," *Human Organization,* XXVI (Winter, 1967), 265–73.

The author of this article focuses on the experiences of Italian immigrants in Australia. The major finding was that the assimilation process tended to be most noticeable when the immigrants settled in neighborhoods not having a large population of other persons of Italian origin. It was further shown that having Australian friends was not absolutely necessary so long as initial motivation to adopt the new lifeways was substantial.

This paper is important not just because it gives a good illustration of the working of the assimilation process but also for the authors examples of concept linkage.

HIMES, JOSEPH S. "The Functions of Racial Conflict," *Social Forces,* XLV (September, 1966), 1–10.

Dr. Himes looks upon racial conflict as a kind of "collective enterprise" aimed at the achievement of "predetermined goals." Looking at the racial situation in the United States from the black point of view he emphasizes the goal of eliminating the pattern of black-white inequality with regard to facilities and opportunities. He sees racial conflict, in light of the goal mentioned, taking the forms of non-violent mass action, political pressure, and legal redress. Dr. Himes believes that, in the sense noted, racial conflict can and does function positively for American society by: (1) altering the social structure; (2) broadening communication; (3) reinforcing solidarity (among blacks especially); and, (4) facilitating personal identity (also primarily for blacks).

PART II—ADDITIONAL READINGS

DYNES, WALLACE. "Education and Tolerance: An Analysis of Interviewing Factors," *Social Forces,* XLVI (September, 1967), 22–23.

LEE, ALFRED MCCLUNG. "The Concept of System," *Social Research,* XXXII (Autumn, 1965), 229–38.

LOOMIS, CHARLES P. "In Praise of Conflict and Its Resolution," *American Sociological Review*, XXXII (December, 1967), 875–90.

PRICE, JOHN A. "The Migration and Adaptation of American Indians to Los Angeles," *Human Organization*, XXVII (Summer, 1968), 168–75.

STRUBING, CARL M. "Some Role Conflicts As Seen by a High School Teacher," *Human Organization*, XXVII (Spring, 1968), 41–42.

WINSBERG, MORTON D. "Jewish Agricultural Colonization in Entre Rios, Argentina; II." *American Journal of Economics and Sociology*, XXVII (October, 1968), 423–28.

4

LEARNING
AND ADJUSTING

culture, socialization,
and personality

In all societies people have to be trained to follow the expectations of their peers. They must be able to fit into desired behavioral niches as adults so they begin to be properly oriented at even very early age levels.

That play could be a useful form of training was understood. Adults encouraged, and supervised to some extent, games in which the children staged "play feasts" and "play pot latches." Even Shaman's Dances were put on by children as a game, with their elders' encouragement and advice.*

*Philip Drucker, *The Northern and Central Nootkan Tribes*, Smithsonian Institution, Bureau of American Ethnology, *Bulletin 144* (Washington, D.C.: United States Government Printing Office, 1951), p. 134.

Requirements For Human Life: Culture

MAN IS BY NATURE a gregarious creature. Since he was not endowed with an armored skin, large fangs and claws, nor enormous strength[1] (despite the fact that *man is one of the largest* of the primates) he has had to band together with his fellows for protection and survival. One result of this group living, which is typical for humans, is that they have developed commonly held modes of belief, patterns of thinking, and ways of doing things in order to cope with the requirements of their environment. Social scientists refer to this conglomeration of mutually shared ideas and actions as *culture*.[2] Since culture is viewed here in its relationship to the environment, the nature of this relationship must be discussed before proceeding further.

Environment and Culture

Any kind of environment[3] presents more than one set of possibilities for human beings to interpret and utilize. While some people believe that environment determines the culture of a social system[4] the evidence for such beliefs is not very good.

[1]Even a chimpanzee, from one-half to three-quarters the weight of an average man, is *much* stronger than a man. See Ernest Hooton, *Man's Poor Relations* (New York: Doubleday Doran, 1942) for a very readable account of the comparative strength capabilities of apes and man. For other man-ape comparisons see John Buettner-Janusch, *Origins of Man: Physical Anthropology* (New York: Wiley, 1966), pp. 350–51.

[2]The basic form of this kind of definition was first formulated by E. B. Tylor in his *Primitive Culture* which was published in 1871. For a comprehensive analysis of the term see A. L. Kroeber and Clyde Kluckhohn, *Culture: a Critical Review of Concepts and Definitions* (New York: Vintage Books, 1952).

[3]Meaning the *total environment*. See Chapter 2.

[4]See, e.g., the works of Arnold Toynbee, a historian, and Ellsworth

Both the Navaho and the Zuni Indians have occupied part of the arid southwestern United States for many centuries, but where the Navaho were once a tribe of semi-nomadic warlike hunters the Zuni have long been relatively nonmobile and peaceful farmers. The different origins of the two Indian groups has nothing to do with the case, for if environment determines culture then any group of people living in a given environment must have very similar ways of life. The fact is, however, that the Navaho and the Zuni each interpret their environment differently. Thus we can only say that the best way to describe the relationship between environment and culture is to say that environment: (1) provides both channels for and limitations on human endeavors, and (2) is not usually perceived the same way by different people.

To take an example, let us suppose that two men, one who belongs to a society where people are taught to swim, and the other from a society whose members are taught to fear water, are being pursued by a lion. If they are chased to the bank of a river there should not be much guesswork involved in determining which of the two is most likely to survive. For one of the men the river will serve nicely to meet his needs but for the other it simply is another obstacle to survival. There is no need to get into a "which came first the chicken or the egg?" kind of argument because we can simply say that culture and environment are interdependent.

Another way of looking at this is to say that the environment setting for any group of people must include the resources to meet their basic needs[5] if the group is to survive. A particular environment may provide resources, such as food and water, but may also be deficient in materials necessary for making clothing or shelter. Of course, if the climate is sufficiently affable, people might be able to get along with relatively little in the way of clothing or permanent shelter, but a decline in the amount of clothes worn may necessitate a rather considerable adjustment

Huntington, a geographer. Note that both these writers actually imply that some groups of people don't do as well in a particular environmental setting as others; thus their environmental determinism boils down to not very well disguised ideas about "racial" superiority and inferiority.

[5]For a discussion of basic needs see Bronislaw Malinowski, *A Scientific Theory of Culture and Other Essays* (Chapel Hill: University of North Carolina Press, 1944), pp. 36–43.

in their whole way of life. In other words, the group will have to make some decision whether to stay in an area that does not provide the wherewithal for clothing, as they presently define it, or they will have to change their definition of clothing, or, perhaps, do without any altogether. These definitions of what should be the ways to respond to needs of the group constitute the components of culture. Culture can be defined as a set of socially defined responses to group needs.[6] These responses are developed in time through relatively continuous social interaction. In short, after members of any society have decided to adopt particular ways of coping with the mandates of their environment they must go about the task of transmitting their ideas to other members of their system. We can see, therefore, that culture is not genetically inherited but must be acquired in interaction with other people.

Thus members of any group may be taught the culture of another. Hence, race and culture cannot be equivalents since persons having similar physical characteristics may belong to culturally different systems while some cultural systems include groups of people representing quite different physical types.

There is some amount of misuse, misunderstanding, and confusion with regard to the terms "culture" and "society." Many persons tend to use the terms as though they are interchangeable when, in fact, they actually have quite different meanings. Society essentially refers to a group of people who, usually, are associated with some geographical area, while culture refers to the behavior commonly observable among some group of people.[7]

Learning To Be Human: Socialization

Socialization refers to the overall process whereby an individual learns a sufficient portion of the culture of his society

---

[6]*Ibid.*, pp. 36–43.

[7]For a comparative discussion of the uses of these terms see Beverly L. Levenson and Myron H. Levenson, "Culture versus Society: A Brief Analytical Overview of Concept Usage," a paper read at the *59th. Annual Meeting* of the American Anthropological Association in Minneapolis, Minnesota, November, 1960. Another paper on the same program dealt with the etymological derivation of the two terms, Roger W. Westcott's, "Society versus Culture: A Study in Terms and Concepts."

(and/or of sub-systems within the society) to enable him to act appropriately in any given situation. In short, the individual new to a social system must learn the patterns of behavior the system has defined as *expected within particular contexts.* Whether the mode of learning takes the form of imitation or of conditioning (responding to reward or punishment) is not so important as the fact that it must take place in association with others. The individual being socialized has to distinguish what behavior is correct for others as well as for himself. The socialization process therefore includes the development of a concept of *self* both apart from and in relation to others.

Not all learning takes place in a social context of course. For example, an individual might, by trial and error, learn how to assemble some mechanical device (such as a clock) without the aid of another person who has the skill to instruct him. However, the learning of *expected behavior* within a social system presupposes that at least one other person who knows the culture will be available to transmit the necessary knowledge.

In everyday discourse we use the terms "I" and "me" to refer respectively to our concepts of ourselves as initiating action or as objects of the action of others. These, however, are simple terms, too simple to adequately describe the totality of one's self-image or how one develops the idea of self. George H. Mead, a philosopher at the University of Chicago, conceived of the development of the ability for "role-taking" as the initial step in socialization.[8] He defined role-taking as the act of putting oneself mentally in the place of another person. He saw that this process involved two things: (1) learning the behavior expected of all persons in a given social context, and (2) developing the capacity to place oneself relative to others within that social context. Mead paid special attention to children at play since, in childhood games, he saw a microcosm of the socialization process.[9] Children, he observed, become accepted as part of a play group only after they learn to act in accordance with the rules of whatever game is being played. Learning the rules includes learning what to expect of each player and, hence,

[8]See George H. Mead (edited by Charles W. Morris), *Mind, Self, and Society* (Chicago: University of Chicago Press), pp. 152–64.
[9]See, e.g., his "Play, the Game, and the Generalized Other," in Anselm L. Strauss, editor, *George Herbert Mead on Social Psychology* (Chicago: University of Chicago Press, 1964), pp. 216–28.

enables one player to think what he would do, the rules allowing, if he were in another player's position.[10]

In a contrasting view Charles H. Cooley, a contemporary of Mead, viewed socialization through the framework of what he called the "looking-glass self."[11] Cooley meant that socialization is a process we undergo all our lives and which operates by means of a person imagining the way in which he is viewed by others. There are three steps in this operation as Cooley saw it.[12]

The first step is the person's imagination of his appearance to other people. In this instance *appearance* refers to the total appearance of a person which includes such things as cleanliness, style and neatness of dress, personal mannerisms, and all overt behavior.

The second step in the process is the imagination of how that appearance is evaluated by others, whether one imagines himself to be seen positively or negatively. Finally there comes the consequent reaction to that imagined evalution—such as pride, shame, or indifference. In short, we are dealing with emotional responses or self-feelings. Cooley clearly saw that over time this process becomes an automatic and even subconscious one but, importantly, he also comprehended that each person develops a concept of himself as both unique and as part of some society.[13]

While Cooley and Mead made some major contributions to the study of socialization, both are essentially "armchair theorists." They did little or no formal empirical research to obtain the kind of data which would support the concepts they developed. Among those who have done such research is the Swiss psychologist Jean Piaget. Piaget has made empirical studies of the socialization process as it operates among children ranging from pre-school age through adolescence.[14] Many of his findings

[10]*Ibid.*, p. 217.
[11]See Charles H. Cooley, *Human Nature and the Social Order* (New York: Scribner's, revised edition, 1930).
[12]*Ibid.*
[13]*Ibid.*, especially pp. 80–160.
[14]See, e.g., Jean Piaget, *The Language and Thought of the Child.* (London: Kegan Paul, Trench, Trubner, 1926); *The Moral Judgment of The Child* (Glencoe, Ill.: Free Press, 1948); *Play, Dreams and Imitation in Childhood* (New York: W. W. Norton, 1951); *The Construction of Reality in the Child* (New York: Basic Books, 1954), Also see Barbel Inhelder and Jean Piaget, *The Growth of Logical Thinking: From Childhood to Adolescence* (London: Routledge and Kegan Paul, 1958).

are cogent to the work of Mead and Cooley but Piag
some important theoretical contributions of his own.
derived from his research are sufficient to illustrate
among students of socialization.

The first of these is the idea that social interaction tah
form of two kinds of relationships: (1) *relations of consi.aint*
involving obligatory rules for behavior and (2) *relations of
cooperation* arising from an awareness of the ideal behavior
standards of a social system.[15]

The second of Piaget's major concepts is that the social self
(or personality) is a combination of inherited psychobiological
tendencies and behavorial patterns learned within the context
of social relationships that occur as part of one's life experiences.[16]

Acting As A Social Unit: Personality

Ralph Linton once made the point that ". . . our only clues
to personality are those provided by the overt behavior of the
individual . . ."[17] In short, we have to observe what people do
and say in order to make estimates of what they are like. Thus,
a definition of personality *as a sociological concept* must include
the notion of observable behavior. Drawing then on the previous
discussion of culture and socialization, and noting Linton's comment, we can define personality as *the open manifestation of
socially acquired patterns of behavior, modified by life experience, and by inherited capabilities and limitations.*[18] More briefly
put, personality refers to individuals' interpretations of the culture of their society.[19]

Since we are saying that culture provides a major part of
the framework of personality and we define culture, in part, as
commonly held behavior patterns within a social system, it stands

---

[15]Jean Piaget, *The Moral Judgment of the Child,* pp. 402–5.

[16]Various works of Piaget, *passim.*

[17]Ralph Linton, *The Cultural Background of Personality* (New York:
Appleton-Century-Crofts, 1945), p. 86.

[18]The core of this definition is drawn from the works of Ralph Linton,
W.I. Thomas, and Florian Znaniecki. *In re* the latter two, see the paperbound
edition of *The Polish Peasant in Europe and America,* (New York: Dover
Publications, 1958). This book should be in the personal library of every
serious student of sociology.

[19]Cf. Melford E. Spiro, "Culture and Personality: The Natural History of a false dichotomy," *Psychiatry,* XIV (February, 1951), 19–46.

to reason that we should expect to find that persons who share similar cultural experiences will reveal some similarities in their personal behavioral characteristics. The term "modal personality" has been used to describe the most commonly observed patterns of behavior in a society.[20] A related concept, derived in part from psychoanalytic theory, is that of "basic personality structure."[21] *Basic personality structure* essentially refers to that portion of the adult personality which has been shaped by common childhood experience in a given cultural setting.[22] An extension of these two concepts, *modal personality* and *basic personality structure*, is to be found in so-called national character studies.[23]

The term "national character" refers to generalized conceptions of the presumed average set of personality patterns which typify the masses of entire large-scale societies.[24] For example, Gorer noted that Americans, more than Europeans, tend to use dating as a source for the building of self-esteem.[25] Dating is seen by Americans as a game in which the object is to build interest in oneself without giving too much to the other party.[26] As Gorer quite concisely puts it:

> It must be repeated that the goal of "dating" is not in the first place sexual satisfaction. An "easy lay" is not a good "date" and vice-versa.[27]

[20]See Anthony F. C. Wallace, *Culture and Personality* (New York: Random House, 1962), pp. 109–11, and John J. Honigmann, *Culture and Personality* (New York: Harper and Bros., 1954), pp. 30–34.

[21]See Abram Kardiner, *The Individual and His Society* (New York: Columbia University Press, 1939), and his "The Concept of Basic Personality Structure as an Operational Tool in the Social Sciences," in Ralph Linton (ed.), *The Science of Man in the World Crisis* (New York: Columbia University Press, 1945), pp. 107–22. Kardiner was trained in the Freudian tradition as a psychiatrist but his capabilities as a social scientist rather than as a clinician far transcend those of Freud who largely viewed all human societies in terms of his experiences with middle class Central Europeans of the Victorian Era.

[22]Cf. Kardiner, *The Individual and His Society*, p. 132–36.

[23]See, e.g., Geoffrey Gorer, *Exploring English Character* (New York: Criterion Books, 1955), or his *The American People* (New York: W. W. Norton, 1948).

[24]For a discussion of the rationale of *national character studies*, see Margaret Mead and Rhoda Metraux, *The Study of Culture At A Distance* (Chicago: University of Chicago Press, 1949). Also, cf., "National Character in the Perspective of the Social Sciences," *The Annals of the American Academy of Political and Social Science*, vol. 370 (March, 1967).

[25]Gorer, *The American People*, pp. 109–19.

[26]*Ibid.*, pp. 109–19.

[27]*Ibid.*, p. 117.

National character research stands in slight contrast with the *modal personality* studies which, while basically the same kind of thing, deal primarily with smaller scale social systems such as a tribe of American Indians or, perhaps a peasant village.[28] *Basic personality structure* can be a guiding concept, implicit or explicit, in either national character or modal personality research since it provides a basis for acquiring comparative data as well as for understanding the effects of family relationships on individual personality development.[29]

For our purposes we will look on personality in the sense of that configuration of socially exhibited behavior considered to be definable in terms of the culture of a given social system. Thus we are taking the viewpoint that *personality represents the behavior of social units.* In other words, we can say that personality is the conceptual link between social system and culture. In the following chapter we will discuss how a socially acceptable personality type is developed.

Problem Questions

1. Using your family as an example of a social system describe three of the important cultural traits which you feel had any strong impact on the development of your personality.

2. Obtain an article or book dealing with the rearing of children in some non-Western society. What points of similarity and difference do you find with your own experiences as a child?

3. What aspects of your present total environment (classroom) do you

[28] E.g., see Ruth Benedict, *Pattern of Culture* (Boston: Houghton Mifflin, 1934); Cora Du Bois, *The People of Alor* (Minneapolis: University of Minnesota Press, 1944); Francis W. Underwood and Irma Honigmann," A Comparison of Socialization and Personality in Two Simple Societies," *American Anthropologist*, XLIX (Oct. 1947), 557–77.

[29] See, e.g., Wayne Dennis, *The Hopi Child* (New York: Appleton-Century-Crofts, 1940); Esther S. Goldfrank, "Socialization, Personality, and the Structure of Pueblo Society (with particular reference to Hopi and Zuni)," *American Anthropologist*, XLVII (Oct. 1945), 516–39; John W. M. Whiting and Irwin L. Child, *Child Training and Personality: A Cross-Cultural Study* (New Haven: Yale University Press, 1953).

see as channels and which do you see as limitations on your capabilities for learning? What then will be necessary to improve your situation?

Selected Adjunct Readings

PART I—ANNOTATED

ALLPORT, GORDON W. "Crises in Normal Personality Development," *Teacher's College Report,* LXVI (December, 1964), 235–41.

In this important study Prof. Allport examined the influence of teachers on the development of personality. He found that there tends to be more effect in mid-adolescence than in early elementary school years. He also noted that a strong influence was perceived by no more than one out of ten students. The particular curriculum itself was not so important as the teacher's mode of presentation. Allport discussed several types of crises and covered the college years in addition to the public school group. This report should be read by all students seriously interested in the process of socialization and personality development.

HATFIELD, JOHN ET AL. "Mother-child Interaction and the Socialization Process," *Child Development,* XXXVIII (June, 1967), 365–414.

This is a study of the interaction of nursery school children with their mothers. Variables were split into two categories: (1) *child variables,* consisting, for example, of dependency, warmth, achievement, aggression, etc.; and, (2) *mother variables,* which include reinforcement, restriction, reward and punishment, etc. Among the important findings were: that there was no great difference in behavior and responses between boys and girls, and, that mothers' behavior was less consistent between sessions than that of the children.

MCCULLERS, JOHN C. and WALTER T. PLANT. "Personality and Social Development: Cultural Influence," *Review of Educational Research,* XXXIV (December, 1964), 500–608.

Essentially this report summarizes research findings which illustrate the effects of both cultural deprivation and cultural enrichment. Among the areas covered were findings which related to implications for personality development among children who were isolated for prolonged periods from social contexts viewed as normal *in terms of white, middle-class standards.* The consequences of college experiences and their impact on personality growth and change was also examined. The most important conclusion of this report is that differences of intelligence, motivation, and personality development are primarily consequences of environmental circumstances.

PART II—ADDITIONAL READINGS

BERKOWITZ, LEONARD. "The Judgmental Process in Personality Functioning," *Psychological Review,* LXVII (March, 1960), 130–42.

COGSWELL, BETTY E. "Some Structural Properties Influencing Socialization," *Administrative Science Quarterly,* XIII (December, 1968), 417–40.

FARBER, MAURICE L. "The Study of National Character: 1955," *Journal of Social Issues,* XI (1955), 52–56.

INKELES, ALEX. "Social Structure and the Socialization of Competence," *Harvard Educational Review,* XXXVI (Summer, 1966), 265–83.

PETERSON, RONALD A. "Rehabilitation of the Culturally Different," *Personnel and Guidance Journal,* XLV (June, 1967), 1001–7.

TANNENBAUM, PERCY H. and JACK M. McLEOD. "On the Measurement of Socialziation," *Public Opinion Quarterly,* XXXI (Spring, 1967), 27–37.

TURNER, RALPH H. "The Problems of Social Dimensions in Personality," *Pacific Sociological Review,* IV (Fall, 1961), 57–62.

WALLACE, ANTHONY F. C. "Individual Differences and Cultural Uniformities," *American Sociological Review,* XVII (December, 1952), 747–50.

5

THE STANDARDIZATION
OF BEHAVIOR

values, attitudes,
norms, and sanctions

Quite often we find it easier to understand social action when we examine the behavior of people who are a bit different from ourselves. As a case in point we can discern, without great difficulty, the major aspects of behavioral standardization in the following:

Young children were scolded or lightly punished. The pulling of an ear or switching on the legs—when they disobeyed, falsified or caused damage in the home. But children above six or seven years of age were expected to know the rules of proper behavior.*

*Verne F. Ray, *Primitive Pragmatists: The Modoc Indians of Northern California* (Seattle: The University of Washington Press, 1963), p. 111.

RALPH LINTON ONCE DEFINED THE ELEMENTS OF CULTURE in terms of socially acceptable and desirable behavior.[1] He presented three main categories of cultural elements: *universals, alternatives, and specialties.*[2] These refer to ideas, habits and responses common respectively to:

1. "... all sane, adult members of the society"
2. a portion of the population but not necessarily all persons belonging to specific groups
3. persons who belong to certain specially designated groupings.[3]

Another way of defining these terms is to say that: *universals* refer to behavior expected of everyone under defined circumstances; *alternatives* refer to a range of permissible behavior in accordance with circumstance; and, *specialties* refer to behavior expected for certain persons in certain narrowly defined instances. As examples of the preceding, let's take the matter of dress and specify an instance for each of Linton's categories. Using an ordinary American as our focus let us suppose that he works in a department store. For our example of a *universal* we can simply say that he *will* put clothes on before going to work. With respect to *alternatives* he may choose to wear a single breasted dark grey wool jacket and black cuffless trousers, or a double breasted dark blue gabardine business suit, or, perhaps a green Edwardian style coat and matching flare bottom pants. If our man, coincidentally, happens to be a floor manager it may be that he will be required to wear a handkerchief bearing the store colors or name in his jacket breast pocket as a mark of his office. The handkerchief would represent a *specialty,* as does his particular occupation.

[1]Ralph Linton, *The Study of Man* (New York: Appleton-Century-Crofts 1936), pp. 271–86.
[2]*Ibid.,* pp. 272–73.
[3]*Ibid.,* pp. 272–73.

Linton's categories are certainly useful, but he neglected something important—those behavioral elements which a society (or a part of one) considers to be undesirable and, perhaps, harmful. We might then add to Linton's list the category of *deviancies* to represent the missing portion of culture. For what a social system considers to be undesirable and unnecessary is as important in defining its culture as those things it considers to be desirable and necessary. Each social system maintains a positive or negative view of different kinds of behavior. That is, its members look upon various behaviors as good or bad, necessary or unnecessary, just or unjust, etc. These behavioral categories are thus evaluated in terms of cultural requirements. Sociologists call any objects, ideas, or actions to which people attach some degree of significance *values.* There are personal values, of course, but we are more concerned here with those things (material or nonmaterial) to which interacting groups of people, social systems, attach particular significance. Let us suppose, for example, that a picnic party has set up its paraphernalia in a meadow. In that meadow let us further suppose that there are a number of small outcroppings of rock. The likelihood is that unless those outcroppings had some unusual features such as bright color or odd-looking formations they would be pretty much ignored by the picnickers. To this picnic party the outcroppings would have no significance; they would not be values. Suppose, however, that a different group of people came to the same meadow—say a team of archaeologists. "Hey look," one of them shouts, "flint!" Flint is an important substance to archaeologists because men of bygone days made many tools from that substance. Hence, whenever archaeologists, specifically those who specialize in prehistoric man, find flint they make some effort to search for the workings of ancient craftsmen—such as spear points, axes, and so on. To those archaeologists flint would have significance—it would be a value. In this case it would be a *positive value* because it is viewed as good or desirable for their purposes. Perhaps to a farmer on whose land such outcroppings of rock occur, however, the flint would be a nuisance which would have to be dug up or plowed around. It would be a *negative value* for him. For different people

---

[4] See page 113 below.

then the very same objects may have different kinds of significance or none whatever.

Let us now examine a different example of values. In some
societies emphasis is put on one or another kind of personal
action, either in relation to a group or on the basis of one's own
initiative. The latter type is a dominant value in the United
States of America where we refer to it as "individualism."
This American value developed in response to frontier conditions
plus the idea of personal freedom which was part of the philosophical structure of the American revolution. In its early form
American individualism was predominantly economic in orientation—everyman as farmer, tradesman, craftsman, trapper, etc.
It has been noted, however, that nowadays individualism is
undergoing a transformation from its original state.[5] Burgess
sees the old economic individualism being transmuted to a cultural form and says:

> The stress now is upon ways of maintaining and en-
> hancing personal development under conditions of modern life.
> The movements of adult education, art appreciation . . . (etc.)
> . . . all emphasize the autonomy of the individual.[6]

Thus, we can say that values may be represented in the form
of material objects or as abstract concepts such as desirable
(or undesirable) orientations for daily behavior. Values can
be looked on then as a set of preferences which direct the attention of people toward certain ends, hence affecting their social
behavior.

The socialization process in any social system also tends to
produce a general homogeneity of attitudes (tendencies to react
in a certain way to designated values) which colors the view
an observer takes of the nature of social behavior, among other
phenomena, within that system.[7] If we say that each social system can be described in terms of a basic set of values and

---

[5]See Ernest W. Burgess, "Social Problems and Social Processes," in
Arnold M. Rose, ed. *Human Behavior and Social Processes* (Boston:
Houghton Mifflin Company, 1962), pp. 381–400.

[6]*Ibid.*, p. 390.

[7]*Ibid.*, and also see Joseph H. Monane, *A Sociology of Human Systems* (New York: Appleton-Century-Crofts, 1967), pp. 31–32.

attitudes we must then face the question of how these values and attitudes are sustained among the bulk of their members.

Social Control: Norms and Sanctions

In order for a social system to maintain its existence as an entity defined by behavioral boundaries the modes of action and attitudes of its members must be relatively well standardized. That is, their overt behavior at least must fall within acceptable limits. Sociologically we refer to such limits as norms. A norm, therefore, refers to *a range of acceptable behavior*. Norms are guides to action—they are *attitude setters*.

We need also to make a distinction between "ideal" and "real" norms. "Ideal norms" may be thought of as what the members of a social system think *ought to be done* while "real norms" refer to *what is most commonly done*, the usual pattern of action. Let's take, for example, the matter of behavior at stop signs. We drivers all know that we *should* stop our automobiles when we come to stop signs, but many times we do not. Still, the majority of us (meaning any percentage over 50) usually do stop, and if not, except in rare cases, we at least slow down rather considerably so that we can safely come to a halt if cross-traffic is noticed—which, after all, is the basic purpose of a stop sign. Thus, while the ideal norm is for every driver to bring his car to a full stop at a stop sign, in many cases the real norm involves merely slowing down sufficiently to be able to come to a complete stop if conditions so dictate. The difference between ideal and real norms leads us to the question of what society does if people do not accept the primary attitudinal focus of a norm? How are the norm violators (meaning those who go beyond all acceptable limits, such as completely ignoring a stop sign) brought into line? Emile Durkheim said that people are constrained to conform because they fear unpleasant consequences which could result from their actions.[8] Certain consequences are the expectable actions members of a social system will take when a norm has been violated. These expectable actions associated with particular norms are called *sanctions*. Sanctions are represented by a complex of rewards (positive sanctions) and punishments (negative sanctions).

[8]See, e.g., Emile Durkheim, *Sociology and Philosophy* (Glencoe, Ill.: Free Press, 1945), pp. 41–45.

Norms may also be viewed as positive or negative in the sense that some refer to actions that should be taken or attitudes that should be held; these are *prescriptive norms*. Conversely, norms that apply to actions that should *not* be taken or attitudes that should *not* be held are *proscriptive norms*.

Positive sanctions are applied when a system deems one or more of its members to be worthy of being rewarded for good or exceptional performance and negative sanctions are applied to those persons whose behavior is considered improper, inadequate, or injurious. Sanctions can take different forms such as[9] (1) *physical types*—applied in some way to the body of a person; (2) *economic types*—applied to a person's property; (3) *psychological types*—applied to a person's feelings or emotions. Positive and negative examples of the preceding are:

1. *physical*:
   a. *positive*—an embrace or a kiss.
   b. *negative*—a kick or a blow to the head.
2. *economic*:
   a. *positive*—a valuable prize or a raise in salary.
   b. *negative*—a fine or confiscation of goods.
3. *psychological*:
   a. *positive*—praise or applause.
   b. *negative*—censure or avoidance.

The exact kinds of sanctions to be applied vary both with the nature of the situation in which the behavior in question takes place and also with the type of social system in which norms are violated or surpassed. Not all social systems have the same norms nor do they all have the same ways of dealing with those who exceed expectations or who fail to behave as desired. In this respect William Graham Sumner noted that what might be right in one society would be wrong in another.[10] Sumner also distinguished between two general types of norms: folkways and mores. He defined folkways essentially as all the traditional ways of conducting ordinary day-to-day activities while mores dealt with matters of life, death, sex, and religion.[11] Mores, then, are norms having to do with matters of serious moral con-

---

[9]Please note that this is not an exhaustive classification of types of sanctions but represents merely a set of common examples.

[10]See William Graham Sumner, *Folkways* (Boston: Ginn and Company, 1906) or Mentor (paperbound ed.) (New York: New American Library, 1940), pp. 17–114.

[11]*Ibid.*, pp. 17–114.

sideration while folkways are norms that have to do with simple matters of appropriateness. In conjunction with this Sumner also made the point that while violation of the *folkways* brings about only mild forms of punishment, any violation of the *mores* of a society usually results in severe sanctions being applied.[12]

We can thus conclude that the members of a society learn to act and think within socially defined limits in order to avoid punishment, and sometimes to gain rewards. What this means is that the vast majority of the members of a particular social system will present many similar patterns of behavior to the observer. Persons who vary any great amount from the combination of patterns common to the majority will be looked upon as odd, if not completely deviant. Therefore it can be said that any society maintains expectations of some degree of behavioral conformity from its members and that it will bring pressure to bear to obtain and to maintain that conformity.

Problem Questions

1. Taking your closest friends and yourself as a small scale social system, what are the two most basic values which characterize the system? What norms (as standards of conduct) are related to those values?

2. Look through some newspapers and locate two articles which discuss sanctions which have been or will be applied to someone or some group of people. What attitudes do you hold about the particular sanctions described? What do you believe are the reasons underlying your attitudes?

Selected Adjunct Readings

PART I—ANNOTATED

DeFleur, Melvin and Frank Westie. "Attitude as a Scientific Concept," *Social Forces*, XLII (October, 1963), 17–30.

[12]*Ibid.*, pp. 17–114.

In this important article the authors discuss the various usages of the concept of attitude. They note that there tend to be two distinct ways of defining the term. One way is referred to as *probability conceptions* wherein consistency of responses toward given stimuli is the core idea. The other mode of usage they call *latent process conceptions* which involve some hypothetical, and not directly observable variable which mediates responses. The authors point out that definitions of the term need to be more closely linked to the methods used to employ them in measurement of social behavior.

GIBBS, JACK P. "Norms: The Problem of Definition and Classification," *American Journal of Sociology*, LXX (March, 1965), 586–94.

Professor Gibbs presents a typology of norms based on collective expectations of behavior, and reactions to behavior. He states that there is no adequate classificatory scheme for distinguishing types of norms. Differences in norms such as mode of enforcement, kinds of sanctions for violations, sources of authority, and amount and kind of conformity are seen to contribute to the confusion in usage of the term. Gibbs sees three basic definitional attributes of norms: (1) collective evaluation of what ought to be, (2) collective expectation of what behavior should be, and (3) reactions to behavior. He notes, however, that extensive research will be needed in order to develop meaningful criteria for collective evaluations and probability of reactions.

WILLHELM, SIDNEY M. "A Reformulation of Social Action Theory," *American Journal of Economics and Sociology*, XXVI (January, 1967), 23–30.

A need for greater precision in defining basic concepts is the focus of this article. Professor Willhelm discusses goals, values, norms, means, and social conditions. His discussion of values is of particular interest. He pays attention to values as a sociological concept by concentrating on four aspects: (1) values in relation to ends by providing specificity, (2) values in relation to means by providing alternatives, (3) values in relation to social conditions in terms of conformity, and (4) values in relation to cognitive data by the imposition of worth or desirability. He goes on to discuss the importance of values in social action in relation to efforts to achieve a particular goal.

PART II—ADDITIONAL READINGS

ANDERSON, ALAN R. and OMAR KHAYYAM MOORE. "The Formal Analysis of Normative Concepts," *American Sociological Review*, XXII (February, 1957), 9–17.

BIDWELL, CHARLES E. "Values, Norms, and the Integration of Complex Social Systems," *Sociological Quarterly,* VII (Spring, 1966), 119–36.

GIBBS, JACK P. "Sanctions," *Social Problems,* XIV (Fall, 1966), 147–59.

LARSON, RICHARD F. and SARA S. SUTKER. "Value Differences and Value Consensus by Socioeconomic Levels," *Social Forces,* XLIV (June, 1966), 563–69.

LEE, DOUGLAS. "The Role of Attitude in Response to Environmental Stress," *Journal of Social Issues,* XXII (October, 1966), 83–91.

McGINNIS, ROBERT. "Campus Values in Mate Selection: A Repeat Study," *Social Forces,* XXXVI (May, 1968), 368–73.

TOBY, JACKSON. "Is Punishment Necessary?" *Journal of Criminal Law, Criminology and Police Science,* LV (September, 1964), 332–37.

YERACARIS, CONSTANTINE A. "Differentials in the Relationship Between Values and Practices in Fertility," *Social Forces,* XXXVIII (December, 1959), 153–58.

6

THE COMPOSITION OF SOCIAL SYSTEMS

social status and
social structure

Americans, as well as citizens of some other nations, like to think of themselves as equals. Everyone is as good as everyone else. In fact, however, distinctions of many sorts are made in all societies. The distinctions result from the social system's evaluations for different characteristics of individuals and groups as in the following case.

The Branchleys are middle class by virtue of their income, housing, and general integrity; they are at the lower end of the class because of poor education, the manual nature of their work, and most of all because they do not participate in community affairs.*

---

*Morton Rubin, *Plantation County* (Chapel Hill: University of North Carolina Press, 1951), p. 120.

EVERY SOCIAL SYSTEM, whether it is a whole society or minute sub-group has some ways of making social distinctions among its members. These distinctions occur in two dimensions, horizontally according to various kinds of categorical variations on the same plane, and vertically, involving judgments of rank. The first dimension, which has been called *situs*,[1] actually amounts to no more than placement in a classifictory grouping which has certain descriptive characteristics. For example, carpenters, persons over 30 years of age, rural-farm dwellers, sultry brunettes, etc., are all kinds of situses. In general, situs may be important to know, insofar as basic characteristics of persons are concerned, but a situs can and often does include persons who may vary rather considerably from one another in many respects, and who may have no relationship other than a statistical one. *Situses* do not carry with them any necessary expectations of the behavior of the persons who compose them and the concept does not then have the social signficance of the second dimension, that of *status*.

Social Status

Status can be defined, in one way, as the level of prestige a person holds according to the various characteristics which describe him as a social unit. In short, *status* refers to a hierarchical system of social placement. A person's status is high, low, or medium in accordance with the way he *in toto*, or in respect to even one distinguishing characteristic, is evaluated by the other members of the social system or systems to which he belongs. Obviously too, a person's status may be, and is more often than

[1] See Emile Benoit-Smullyan, "Status Types and Status Interrelations," *American Sociological Review*, IX (April, 1944), 151–61.

not, judged in comparison with that of others, which thus brings the ideas of both reciprocality and superiority-inferiority into play. As a general rule it can be said that the higher a person's status the greater the number of rights and privileges he enjoys. Since the concept of status automatically implies *comparative placement within a social system* it must be a truly social variable and can not then be applied to an individual who is not being viewed as a social unit.

E.T. Hiller has developed the concept of *key status* which means the overall or average status a person holds in his community according to the way all of his status characteristics are judged by his fellows.[2] Ordinarily key status is associated with a person's most important function in his community, which usually is his occupation,[3] but there are other factors which are not necessarily behavioral in nature that can intervene, such as (1) physical and mental endowment (including age, race, and sex), (2) education, training, and experience (not necessarily the kind relating *only* to one's occupation), (3) interests and preferences (including religion and politics as well as one's choice of friends and associates), and (4) ethnic background (that is, the cultural group of origin of one's ancestors). To take an example of what is meant here, a black craftsman, *no matter how skilled,* is likely to have a lower status in many American communities than his white counterpart simply because he is black. Thus, even those characteristics we cannot help being born with such as race, sex, eye-color, etc., may have an important bearing on our placement in the status ladder of a given social system. It would not be amiss to say that any single characteristic (or combination of them) of any kind which can be used to describe human beings may serve as a determinant of status.

For example, it is perfectly feasible to randomly select one of the more complex tables from the reports of the United States Census of Population and utilize the variables stated in the table headings for a study of status effects. If we were to take as our example Table 103 from the 1960 census report on the population of North Carolina we will find that the table is titled

[2]See E. T. Hiller, *Social Relations and Structures* (New York: Harper and Bros., 1947), 339–43.
[3]*Ibid.*, pp. 339–43.

as "Years of School Completed by Persons 14 years old and over, by age, color, and sex, for the state, 1960 and 1950, for Urban and Rural Areas, and for Standard Metropolitan Statistical Areas of 250,000 or more, 1960."[4] Any of the variables mentioned in the table title can be used as a measure of status by itself or in combination with any of the others. Thus, a researcher might choose to concentrate on years of school completed, age, and sex as his status factors, or he might choose to select years of school completed, color, and urban or rural residences, perhaps, in the latter case, using urban residence as the higher ranked status element.

Because status is dependent on the cultural values of a society and the views people take of one another in accordance with those values (and their own biases) we could say that status is primarily symbolic in nature. This is not to say that status is by any means imaginary; we can see too many examples of it in everyday life to believe that. On the contrary, it is because status is a *socio-psychological reality* that we must pay special attention to it. People *want* status, they will strive for more and better rights and privileges and will do their best to avoid losing prestige.[5] Thus any change in status may be seen as something attractive or fearful according to a person's conception of what is a minimally acceptable level of prestige for his present and presumed future place in society.[6]

While status may be expanded or contracted in certain instances according to how well or how poorly a person performs in various social situations some status levels are inherently fixed within particular social systems. The matter of the black craftsman was mentioned before and we can also note that the simple matter of reaching a certain age is sufficient to confer a given status on people. For example, once one becomes twenty-one years of age in the United States many doors are opened to him which might previously have been closed without parental permission—most states maintain twenty-one as the minimum

[4]See U.S. Bureau of the Census, *U.S. Census of Population: 1960 Detailed Characteristics, North Carolina*. Final Report PCC123510 (U.S. Government Printing Office, Washington, D.C., 1962), pp. 356–60.

[5]Cf., Alexander Leighton, *My Name is Legion* (New York: Basic Books, 1959), pp. 145–46.

[6]*Ibid*. Also see Helen Flanders Dunbar, *Psychiatry in the Medical Specialties* (New York: McGraw-Hill, 1959), p. 10, and *Mind and Body: Psychosomatic Medicine* (New York: Random House, 1947), p. 246.

age of marriage without consent of parents, and in most states it is the minimum legal age for purchase of alcoholic beverages.

Naturally groups of people have their desires and their conceptions of desirable and undesirable statuses as well as individuals. Groups are also accorded greater or lesser levels of prestige by their societies. It is when we look at status differences in terms of levels of status for entire social systems that we can better examine the overall effects of status on human society.

Social Structure

The structure of a social system can be defined as a network of relationships.[7] If these relationships between members of a system were all identical we would not need to go any further than to describe any one of them. However, these relationships are not all identical since they involve interaction between and among persons who are seen as having different amounts of prestige and consequently a greater or lesser quantity and quality of rights and privileges.

We can then say that social structure consists of a *network of statuses*. The statuses range from high to low and various groups tend to fall within a particular range of status levels. We call this hierarchical patterning of status levels *social stratification*. Social stratification, as a major pattern of social structure, takes a variety of forms and we can classify those forms in three categories with respect to the relative degree of social mobility that is associated with each. *Social mobility* refers to the movement possible between and within social strata. We will not be particularly concerned here with changes within strata, such as a shift from one occupation to another, that is *horizontal* mobility, but rather we will pay special attention to the question of ability to move up or down the scale of status, which is *vertical* social mobility. For our purposes we will classify stratification orders as (1) *closed*—characterized by a complete lack of mobility, (2) *restricted*—characterized by partial mobility, and (3) *open*—meaning that a high degree of mobility is possible.

---

[7]Cf., A. R. Radcliffe-Brown, "On Social Structure," in his *Structure and Function in Primitive Society* (Glencoe, Ill.: Free Press, 1952), pp. 188–204 and especially pp. 191–92.

The major examples of each of these orders are usually discussed as *caste, feudal,* and *class* patterns of stratification.[8]

The term *caste* comes from the Portuguese *casta,* meaning a box with a tightly fitted lid. Once something is placed within such a box it cannot easily be removed. This was the view the Portuguese took of the caste stratification system they found in India. They found that people could not move up or down the scale of status levels nor could they very easily make shifts within a particular caste level. Thus, social mobility is not a characteristic of the caste pattern of stratification.

In the middle Ages in Europe a way of life which has been termed *feudal* came into existence. This way of life or culture had its own characteristic pattern of stratification. At the top, of course, were the nobility who owned the land and everything on it. At the bottom were the serfs who tilled the soil and cared for the livestock. In between were the few freeman who acted as skilled craftsmen, merchants, and soldiers. On rare occasions the nobility *might* marry one of the middle group, sometimes lowering their own status in the process, but perhaps more often raising the status of the middle group person. On even more rare occasions a serf might act valiantly in defense of his master or his master's flocks and be elevated to the level of a freeman. But such occasions were indeed rare. Thus we can say that the feudal stratification pattern is characterized by a *very limited* degree of social mobility.

Unlike the caste and feudal patterns *class* stratification systems theoretically have no barriers to mobility. People can move up and down the status ladder and change, horizontally, their situs locations as well. Still the reality is that the ability to be mobile in a class system is dependent on a number of factors. Among these are the basic considerations of education and ability, sex and ethnic background, income, etc. The class system, as we know it today, is a relatively late development. The caste pattern of India goes back a few thousand years. The feudal pattern, in what can be called its beginning stages is probably almost as old, if not equally so, as the caste pattern.[9] The class pattern

[8]Cf., Egon E. Bergel, *Social Stratification* (New York: McGraw-Hill, 1962).

[9]Cf., Harold M. Hodges, Jr., *Social Stratification: Class in America* (Cambridge, Mass.: Schenkman Publishing Co., 1964), pp. 17–38.

of stratification, however, possibly cannot be traced back much farther than the Renaissance.[10] With regard to social class as presently defined we need go back no further than the late eighteenth and early nineteenth centuries to find the concept discussed in the works of such men as Adam Ferguson or Henri, Comte de Saint-Simon.[11] However, for our purposes, we will jump to the twentieth century for examples of representative theorists.

One of the major sociologists writing about social class in this century was Max Weber. Weber believed class stratification should be viewed in terms of three elements: *class, status,* and *party*.[12] He defined *class* essentially as position on the economic ladder. *Status* referred to the prestige element concerning personal achievements and group distinctions, and *party* meant place within the power structure of a society.[13] Unlike Karl Marx who conceived of social structure primarily in economic terms, with the economy determining the form of all the rest of society's structure,[14] Weber was able to see that economic position alone was inadequate for explaining social stratification.[15] Weber, however, had an advantage over Marx by his later arrival

---

[10]There will undoubtedly be some who will argue for earlier or later periods but the main point here is that the class system, regardless of time of origin, is the youngest of the three types so far as can presently be determined.

[11]Adam Ferguson, *Essay on the History of Civil Society* (London: 1767). Henri, Comte de Saint-Simon published a number of works between 1803 and 1825. Among his notable ones are *Introduction aux Travaux Scientifiques du XIX Siecle* (Introduction to Scientific Works of the 19th Century, published in 1807–1808), *De La Reorganization de La Societe Europeene* (On the Reorganization of European Society, published in 1814), *Du Systeme Industriel* (On the Industrial System, published in 1821–1822), and *De l'Organization Sociale* (On Social Organization, published in 1825).

[12]See H. H. Gerth and C. Wright Mills, eds. and trans., *From Max Weber: Essays in Sociology* (New York: Oxford University Press, 1958), pp. 180–95. This excerpt was taken from Weber's *Wirtschaft and Gesellschaft* [Economy and Society], an uncompleted work from which a number of sections have been published and translated. The first section was translated into English by A. M. Henderson and Talcott Parsons with the latter also editing the work which is called *Max Weber: The Theory of Social and Economic Organization* (New York: Oxford University Press, 1947).

[13]Gerth and Mills, *From Max Weber: Essays in Sociology*, pp. 180–95.

[14]Karl Marx, *Capital* (edited by Frederick Engels and translated into English by Ernest Untermann) (Chicago: Kerr Publishing Co., 1909).

[15]Gerth and Mills, *From Max Weber: Essays in Sociology*, and Henderson and Parsons, *Max Weber: The Theory of Social and Economic Organization*.

on the scene and the greater depth and breadth of his historical research. Some substantiation of Weber's ideas came in the 1930s and 1940s as a result of a number of studies made on class and stratification. One notable example among the researchers of that time was W. Lloyd Warner. Warner, an anthropologist who had worked among the Australian aborigines, decided to turn his talents to an examination of the structure of American society. He, and a host of associates, conducted the now famous "Yankee City" series of studies[16] which established a pattern for research on class structure that became common among American sociologists. Warner and his associates found that much more than economic position was necessary if one intended to explain stratification. They found, among other things, that religious affiliation, level of education, and kinds of organizational memberships were important in determining one's class position.[17] They also are famous for their sexpartite grouping of the American social strata into: upper-upper, lower-upper, upper-middle, lower-middle, upper-lower, and lower-lower social classes.[18]

While the findings of Warner and his group help to some extent to support Max Weber's ideas, an important part of Weber's thesis is left out in their considerations. Not ignored, but rather presented more by implication than specific declaration, is Weber's idea that position in the power structure of a society ("Party") is an important aspect of stratification.[19] We may note here that the Warner team developed two basic approaches to the study of American social structure. The first, their "objective" approach, took the form of what they called the *Index of Status Characteristics*.[20] This consists of determining (1) amount and sources of income and (2) place and type of residence. Their second approach, the "subjective" one, they called *Evaluated Participation*.[21] This consists of forming a panel of judges made up of representative members of a com-

---

[16]See, e.g., W. Lloyd Warner and Paul S. Lunt, *The Social Life of a Modern Community* (New Haven: Yale University Press, 1941).
      [17]*Ibid.* Also see W. Lloyd Warner, Marcia Meeker, and Kenneth Eells, *Social Class in America: A Manual for Procedure for the Measurement of Social Status*, (Chicago: Science Research Associates, 1949).
      [18]Warner and Lunt, *op. cit.*
      [19]Warner and Lunt, *op cit.*; Warner, Meeker, and Eells, *op. cit.*
      [20]Warner, Meeker, and Eells, *op. cit.*
      [21]*Ibid.*

munity and asking them to place families and individuals of that community in groups ranked from high to low status; their judgment being based on: (1) known accomplishments, (2) the organizations each belongs to, (3) their known associates, and (4) general reputation. Power position may play a part in the latter procedure but it is *not* a primary element in the Warnerian scheme of analysis. Yet *power*, the ability by any means (physical, economic, or psychological), to control the behavior of others, is an obviously important facet in the placement of persons within the structure of a social system.

The concept of power should not be confused with *authority* since the latter term has the meaning of the legitimate right to use power, whether or not any real power of any type exists in the individuals or groups concerned. For example, if one person has a gun and some others don't, the former has power, but not necessarily authority, to use his weapon against the others. If, however, the gun-bearer is a duly constituted law enforcement officer and the others an unruly mob he may have the authority to use his weapon under given circumstances, for instance, if he is attacked. On the other hand a person or group can have authority without power as in the case of our federal Food and Drug Administration which has been hampered by limitations on its power for many years. There is no law on the books which requires a food processor to prove that any additives, whether preservatives, coloring agents, or artificial flavorings which are put in the food are fully safe for human consumption!

Back in the first quarter of the 20th century the Italian scientist-philosopher Vilfredo Pareto had noted that in all societies social relationships can be examined in terms of position in the power structure. He saw that in all social systems men were either leaders or followers (in respect to other men) and that changes in the social structure usually took the form of "circulation of the elite." He meant by this that occasionally people at the top of the power structure are pushed out and that occasionally a few from the bottom rise up.[22] These facts must be accounted for and, hence, we need to comprehend the power structure of a society if we are to understand its system of social stratification.

A number of attempts have been made to analyze the power

---

[22]See, e.g., S. E. Finer, *The Sociological Writings of Vilfredo Pareto* (New York: Praeger, 1966).

structure in the modern United States. Good examples of such studies have been the works of Floyd Hunter, C. Wright Mills, and Suzanne Keller.[23] Hunter wrote about the top leaders of a Southern city and noted the pervasiveness of their influences throughout a sizeable area. Mills discussed the place of economic, political, and military elites for the United States as a whole, but neglected the influence of labor leaders and religious figures. Keller analyzed the influence of strategically placed people in all kinds of societies. While these are valuable additions to our knowledge and theory about the power aspects of stratification they, by themselves, cannot be utilized to comprehend stratification. The fact of the matter is that whether one looks at stratification from the economic standpoint or the power standpoint either approach is inseparable from the status concept. Economic gain can build status but high status may aid in securing economic gain! The same relationship obtains between power and status so we are back to *status as the key concept in stratification*. To sum up briefly, we can say that social classes are made up of people who fall within particular ranges of status *regardless of how their individual status levels were attained*.

Functional and Invidious Stratification

In the examination of stratification it is important to note that there are two major facets of stratification which must be considered. The first is that social ranking may be seen to have desirable consequences for a social system so that particular stratification patterns will be built into the culture as part of the set of expected relationships among system members. That is, the members will be taught to have certain expectations of status for certain people. Some persons and groups of persons

[23]See Floyd Hunter, *Community Power Structure* (Chapel Hill: University of North Carolina Press, 1953); C. Wright Mills, *The Power Elite* (New York: Oxford University Press, 1956); Suzanne Keller, *Beyond the Ruling Class: Strategic Elites in Modern Society* (New York: Random House, 1963).

will be perceived as higher or lower on the status scale than others and consequently some will be expected to have more rights, privileges, and power than others.

The second facet is that in some systems there will be people who attempt to ensure high status levels for themselves by denying it to others. The reasons and rationalizations for this invidious discrimination vary but the persons to be denied their rights or restricted in privilege will always have some characteristics, physical, behavioral, or cultural, which the others can declare to be bad, incompetent, or untrustworthy. Once again our example of the black craftsman is a good case in point.

The fact that a person can rate high on the prestige scale due to his abilities (or other characteristics) and yet low on the prestige scale because of his race (or any other characteristics) indicates that there is a condition of *status inconsistency* which can exist in any social system. Such inconsistency often comes into being when functional and invidious aspects of stratification clash with one another, and may produce distinctive forms of behavior. Goffman,[24] for example, has found some relationship to exist between a lack of consistency among statuses and a desire for substantial change in the American political system. Jackson,[25] in a different kind of study, has found status inconsistency to be related to psychological stress. This combination of findings could lead one to the conclusion that status consistency, or its lack, should be an important topic for sociological research. In fact, a not inconsiderable amount of attention has been and is being paid to research on the topic.[26]

There are, of course, other topics in the field of social structure and status relationships which are of some importance to social scientists.

Davis and Moore have discussed stratification as a necessary part of the structure of any society in the sense that those

[24]Erwin W. Goffman, "Status Consistency and Preference for Change in Power Distribution," *American Sociological Review*, XXII (June, 1957), 275–81.

[25]Elton F. Jackson, "Status Consistency and Symptoms of Stress," *American Sociological Review*, XXVII (August, 1962), pp. 469–80.

[26]See, e.g., Nico Stehr, "Status Consistency: The Theoretical Concept and its Empirical Referent," *Pacific Sociological Review*, XI (Fall, 1966), 95–97.

social positions defined as most important for the operation of the society will be the best rewarded.[27] Thus their view is that status inequality is a normal characteristic of human society. Admittedly such a position is limited in its ability to explain all forms of stratification and for this they have been roundly criticized.[28] On the other hand, Davis and Moore have received some support from Talcott Parsons. Parsons, noting that stratification patterns are related to the values of a society,[29] emphasizes that insofar as some persons will have authority over others in any social system they will then have *superiority* to them. This means, in certain contexts, that they have higher status.[30] While it is possible to accept the arguments of Davis and Moore, and Parsons, with regard to the necessary aspects of stratification, their ideas do not affect the fact that stratification may not always be useful in a particular form or that some forms may never have to exist as a part of the social structure of an entire society. Let us assume that stratification based on differences of importance of social positions and on required levels of authority is a must for *all* social systems. We can discern, however, that some stratification patterns exist which have nothing to do with either factor. For example, stratification based on racial or ethnic origins can have no purpose except for that of deliberate discrimination against specific groups of people. In a discussion of ethnic and racial barriers to social mobility Kahl mentions, for example, that:

> . . . because Negroes are lower in the scale, they are thought to be fundamentally inferior, and thus incapable of rising. This makes it harder for them to rise, they are caught in a vicious circle.[31]

[27]See Kingsley Davis and Wilbert E. Moore, "Some Principles of Stratification," *American Sociological Review*, X (April, 1945), 242–49.

[28]See, e.g., Melvin Tumin, "Some Principles of Stratification: A Critical Analysis," *American Sociological Review*, XVIII (August, 1953), 387–94; and Richard Simpson, "A Modification of the Functional Theory of Stratification," *Social Forces*, XXXV (December, 1956), 132–37.

[29]Talcott Parsons, "A Revised Analytical Approach to the Theory of Social Stratification," in Reinhart Bendix and Seymour M. Lipset (eds), *Class, Status and Power: A Reader in Social Stratification* (Glencoe, Ill.: Free Press, 1953), pp. 92–128.

[30]*Ibid.*, pp. 107–8.

[31]Joseph A. Kahl, *The American Class Structure* (New York: Rinehart and Company, 1957), p. 227.

Thus, the American Blacks are kept in an inferior status because they are believed to be inferior, hence, the only thing to do is to make sure that they are kept inferior! Such status differentiation is clearly based on prejudice and rather than being functional for American society it is dysfunctional since invidious discrimination is against the basic American value system and requires the public to pay for the long term results of such discrimination in the form of taxes for welfare, special training facilities, riot control, etc.

We can see status differentials operating in a much more rational and useful fashion when we look at the relationships between, say, foremen and production line workers, aircraft pilots and other air crew members, or, perhaps, research scientists and laboratory technicians. This kind of stratification is related to the division of labor in a society and is indicative of the close relationship that exists between status and the activities in which people engage. We shall note that relationship and other important facets of social life in the next chapter.

Problem Questions

1. Consider the student body of which you are a member as a stratified social system. What are the major status factors? Does invidious stratification exist *within* the student body, and if so, what are its characteristics and how do you explain its existence? Looking at your campus as a whole what kind of stratification order (caste, feudal, or class) if any, exists among the administration, faculty, and students when they are looked on as a society? Justify your answer with facts readily available to anyone.

2. Examine one of the novels written by Sinclair Lewis, John P. Marquand, or Frank Norris.* What picture of American class structure do you perceive? Do you believe such patterns exist today? If so, give factual reasons for your belief from newspaper reports.

*As examples, you might consider, respectively: *Babbitt, Point of No Return,* or *The Octopus.*

Selected Adjunct Readings

PART I—ANNOTATED

AXON, GORDON. "Sociology of American Class Distinctions," *Contemporary Review*, CCX (June, 1967), 307–9.

Mr. Axon finds that Americans tend to refer to one another in terms reflecting class distinctions. He notes that people in the political middle don't become excited about public issues until they see their interests being threatened. He believes that generally very little progress is being made in American society because most of those who have achieved a relative degree of prosperity are antithetical toward those who have not. He sees American class distinctions as being based on wealth and education with the average American being generally unable or unwilling to do much about social injustice. Whether or not one agrees with Mr. Axon's basically Marxist position, he presents a highly interesting and thought-provoking view of modern American social structure.

GITTELL, MARILYN. "A Typology of Power for Measuring Social Change," *American Behavioral Scientist*, IX (April, 1966), 23–28.

This author presents one of the best and most concise analyses of power structure available. She utilizes four power elite models which, in effect, represent four different kinds of social structure. One is the *single elite* type where there is no competition and change is unlikely. The second is that where a *primary elite* is challenged by a *secondary elite* and changes are essentially limited to protecting the *status quo* of important power elements. Her third type is the case of *power diffused* with many *equal competing* elites and any changes tend to produce more power for some groups and less for others. The final type is the *vertical elite structure* where two or more strata of elites compete and change is gradual and widely distributed among the power elements of a social system. This is a fine analytical article which should be read by any student interested in the linkage between social structure and system change.

KRAUSS, IRVING. "Some Perspectives on Social Stratification and Social Class," *Sociological Review*, XV (July, 1967), 129–40.

Professor Krauss provides the reader with an interesting distinction between *social stratification,* which he views as a status system, and *social class,* which he sees as representing particular groups and their characteristic behavior. Focusing on class conflict the author points out that in those societies where status consistency is greater, the amount of class conflicts is also greater. He explains this in terms of the more clearly outlined positions for each class combined with limitations on social mobility. This is an important article for the student of status and class relationships.

PART II—ADDITIONAL READINGS

BOCK, PHILIP K. "Three Descriptive Models of Social Structure," *Philosophy of Science,* XXXIV (June, 1967), 168–74.

D'ANTONIO, WILLIAM V. and JULIAN SAMORA. "Occupational Differentiation in Four Southwestern Communities: A Study of Ethnic Differential in Hospitals," *Social Forces,* XLI (October, 1962), 17–26.

HODGE, ROBERT W. and DONALD J. TREIMAN. "Class Identification in the United States," *American Journal of Sociology,* LXXIII (March, 1968), 535–47.

JEFFREY, C. RAY. "Social Class and Adoption Petitioners," *Social Problems,* IX (Spring, 1962), 354–59.

LANDIS, JUDSON R. et al. "Race and Social Class As A Determinant of Social Distance," *Sociology and Social Research,* LI (October, 1966), 78–87.

LOPREATO, JOSEPH. "Social Classes in an Italian Farm Village," *Rural Sociology,* XXVI (September, 1961), 266–82.

OLSEN, MARVIN. "Percieved Legitimacy of Social Protest Action," *Social Problems,* XV (Autumn, 1968), 297–310.

ROBINS, LEE W. et al. "The Interaction of Social Class and Deviant Behavior," *American Sociological Review,* XXVII (August, 1963), 219–29.

WILLIAMSON, ROBERT C. "Some Variables of Middle and Lower Class in Two Central American Cities," *Social Forces,* XLI (December, 1962), 195–208.

7

THE ORGANIZATION OF HUMAN ACTIVITY

social roles and
social institutions

Every person in every social system has particular activities in which he must engage. These activities are, in part, basic to the values of the system to which he belongs.

. . . the men devote much time during the fall months in preparation for winter. Sleds are built or repaired and dog harnesses are made ready.*

*James W. Vanstone, *Point Hope: An Eskimo Village in Transition* (Seattle: University of Washington Press, 1962), p. 30.

Social Roles

THE ACTIVITIES in which members of a social system engage are a major part of their sociocultural environment. For each activity there is a socially developed set of behavior patterns expected of each participant. These expected behavior patterns are called *roles.* Roles can be viewed as behavioral situses but not all situses are roles. The role of a policeman, for example, is also a situs. A policeman might have red hair and for that reason will be in a situs with other red haired persons, some of whom (or even all) will have few, if any, behavioral characteristics in common.

Red hair has nothing to do with the behavior expected of a policeman. Being a policeman has no necessary connotation of the color of a person's hair. Being red haired and being a policeman constitute two different kinds of situses of which *only the latter* is a role. Additionally the fact of being a policeman or having red hair or both will have some effect on a person's status in his community depending on how the fact of being red haired or being on the police force happen to be valued in that community.

If we keep in mind that *situs* represents only membership in a descriptive category, regardless of the nature of the descriptive characteristics employed, while *role* always implies a set of behavorial expectations, the distinction can be more easily comprehended. Another way of describing the relationship between role, situs, and status is to use verbal shorthand and merely say that *persons enact roles, occupy situses,* and *possess statuses.*

To be sure, a status category is also a situs category but this merely indicates the over-generality of the situs concept, as mentioned in the previous chapter.[1]

---

[1] See above, pp. 58–59.

We must also consider that every person plays a number of different roles in his society and each role carries with it an expected level of performance. Roles must be enacted properly to insure social acceptance and to avoid negative sanctions. In other words people are judged by others not only for what they do but also for how they do it. This question of role performance is a central one in the analysis of status. Every social role carries with it a particular level of status in terms of the prestige and privileges associated with the role. Furthermore, the kind of performance presented by the role player is also a status factor. A good performance assures a somewhat higher status level than an average performance, and a poor performance will result in a correspondingly lower status than average for persons enacting that role. We can say that while status implies prestige and privilege, *role* implies duties and obligations.[2] The duties are the actions which are peculiar to a given role while the obligations are of two kinds: (1) that the actor does certain things and (2) that he does them in some socially preferable manner.

Each activity in a social system is made up of a number of roles which can be classified in two categories: those which comprise certain actions initiated by a role player and those which involve actions that are conducted in response to those of others.[3] If we focus our attentions upon a particular role we will find that it stands in a special relationship to other roles. That is, for each role there is one or more other roles which are complementary to it. In a sense, then, social roles exist only in combination with other roles. By "other roles" is not meant additional roles for a given actor but rather those roles played in relation to his. This pattern of related roles is referred to as a *role-set.*[4] Consider, for example, a married man. He plays several roles in his family. He enacts the role of husband vis-à-vis his wife; hence, the husband and wife roles constitute one

[2]Cf. Ralph Linton, *The Study of Man* (New York: Appleton-Century-Crofts, Inc., 1936), pp. 113–31.

[3]See Talcott Parsons et al. "Some Fundamental Categories of the Theory of Action: A General Statement," in Talcott Parsons and Edward A. Shils (eds), *Toward a General Theory of Action* (Cambridge, Mass.: Harvard University Press, 1951), especially pp. 4–5.

[4]Cf., Robert K. Merton, "Continuities in the Theory of Reference Groups and Social Structure," in his *Social Theory and Social Structure,* revised edition (Glencoe, Ill.: Free Press, 1957), pp. 367–80.

role-set of that family. With regard to his child the same man plays the role of father which means that *father and child* represent another family role-set.

Finally, when we consider that roles contain certain expectations of behavior, some acts being required, others permitted, and some forbidden we can also say that roles represent norm complexes.[5] Roles, then, are behavioral and normative elements of a social system.

Social Organization and Institutions

Whereas the network of statuses represents the social structure of a society the network of roles comprising the activities of the society represents its social organization. Social structure is a picture of the relationship patterns of the people who make up a social system. Social organization, however, means the picture of the system in operation because it reveals how the various activities interconnect.[6]

In order to meet group needs man long ago began to develop patterns of activities and associated roles which together formed distinctive complexes related to particular aspects of the total environment with which he had to cope. Over time these activity-role complexes became more or less standardized; that is, they became sets of group norms cohering about a given area of need. We call such standardized norm-sets *institutions.* Another way of defining institutions is to say that they represent patterns of values and relationships which regulate behavior directed toward meeting social system needs.

Among the institutions common to most societies are one form or another of the following:[7]

1. *Kinship*—which regulates sex behavior from courtship to procreation, and relationships among genetically and affinally connected persons (family).

[5] Cf. Frederick L. Bates, "Position, Role and Status: A reformulation of Concepts," *Social Forces*, XXXIV (May, 1956), 313–21.
[6] Cf. A. R. Radcliffe-Brown, *Structure and Function in Primitive Society* (Glencoe, Ill.: Free Press, 1952), p. 11.
[7] There are, of course, common institutional forms other than those listed here as examples: See, e.g., Joyce O. Hertzler, *American Social Institutions* (Boston: Allyn and Bacon, 1961), and Constantine Panunzio, *Major Social Institutions* (New York: Macmillan, 1939).

2. *Economy*—which regulates behavior directed toward the acquisition and distribution of food and other necessities for sustaining life.
3. *Education*—which regulates behavior involved in the acquisition and transmission of knowledge, skills, and basic values of a society.
4. *Polity*—which regulates behavior in general and acts to maintain societal order.
5. *Religion*—which regulates behavior related to social concepts of man's place in nature and those phenomena beyond his immediate understanding.
6. *Art*—which regulates creative behavior and self-expression through the definition of aesthetic values.
7. *Recreation*—which regulates behavior that contributes to the refreshment of human mental and physical conditions.
8. *Protection*—which regulates behavior oriented toward defense against hazards of the environment.

Naturally the institutional forms described above are very general in nature. Subtypes of each can be, and often are, emphasized instead of the more general and inclusive kinds.[8] One could, for example, denote the military and medicine as sub-types of the institutional area of protection.

The important thing to keep in mind about institutions is that they represent the organization of activities in a social system. In so doing they also stand for certain values and imply, if not define, particular kinds of status and those persons who, enacting necessary roles, will hold given status levels. Consequently, if we describe all the basic institutions of a society, we are describing its culture.[9]

We can say that *institution* is an abstract concept because we cannot see institutions. However, anthropologist Bronislaw Malinowski has provided us with a definition of an institution, not as a totally abstract entity, but rather as an operative mechanism for meeting needs. Malinowski said that an institution consists of the following empirically determinable components:[10]

1. *A charter*—the socially defined statement of purpose of an institution—its *raison d'etre*.

---

[8]See, e.g., Gerard De Gre, *Science As A Social Institution* (New York: Random House, 1952). Science, for example, is part of the larger institution of education.
[9]See Bronislaw Malinowski, *A Scientific Theory of Culture and Other Essays* (Chapel Hill: University of North Carolina Press, 1944), p. 40.
[10]*Ibid.*, pp. 52–54.

2. *Personnel*—the people who carry out the functions of the institution plus those who are served by it at any given time.
3. *Rules of behavior*—the norms which guide the actions of persons involved in meeting institutional goals.
4. *Aims and activities*—specific ends to be attained and prescribed ways of attaining them.
5. *Apparatus*—material items to be used in conducting institutional activities or to represent the institution. The former are called *utilitarian* apparatus and the latter are referred to as *symbolic* apparatus.

To avoid confusion between institution as an abstract concept and institution in the concrete sense described by Malinowski we can call the latter *Institutional facilities*. An example of institution as an abstract concept would be the economy while a similar example of institution in the concrete sense would be a bank. A bank is a *part* of the total economic structure of a society. The general pattern of values which would describe the nature of the economic structure is an abstraction, for instance, the economy of the United States.

The term "association" is sometimes used as a synonym for "institutional facility," and so is the term "organization." I think "facility" is the more preferable term, however, since *association* is often used variously to mean everything from more or less casually formed recreational groups ("Mom's bridge club," for example) to highly organized systems such as military units, industrial complexes, or national governments. The term "organization," on the other hand, can easily be confused with the process of forming and coordinating a group or its activities.

Institutionalization, Bureaucracy, and Bureaucratization

Institutions, and the facilities which serve to attain their goals, can be looked on as patterns of formalization of behavior. The process of such formalization is called *institutionalization*. Institutionalization essentially consists of the establishment, regulation, and coordination of attitudes and actions which have the purpose of meeting some needs of a given social system. In large scale systems, where there is a necessity to guide and control the behavior of a large number of people engaged in many different and often very complex activities the institutional facilities tend to take the form of bureaucracies.

A *bureaucracy* is not a group of people but rather is a form of administration. Max Weber discussed the characteristics of a bureaucracy in some detail.[11] He noted, among other things, that bureaucratic organization consists of:

1. A complex hierarchy of levels of authority.
2. Specified areas of jurisdiction for each administrator.
3. Clearly defined tasks and procedures for accomplishing them.
4. Selection of officials on the basis of qualifications needed for particular tasks.[12]

To many people the term "bureaucracy" connotes negative ideas such as inefficiency, mismanagement, delays, or rising costs. However, those ideas apply only to bureaucratic facilities which are not operated properly and not to all bureaucracies at all times.

Weber essentially defined bureaucracy as an efficient system of administration for large scale groups no matter what kind of goals happen to be pursued by them.[13] The process of converting the operation of large scale social systems into a bureaucratic form is called *bureaucratization.* Bureaucratization is a special case of institutionalization. In this case the intent is to regulate and coordinate actions and attitudes of a large number of persons engaged in a network of interrelated tasks aimed at attaining particular goals.[14]

While the basic purpose of a bureaucracy is to promote efficiency the results do not, unfortunately, often live up to expectations. Among the many reasons for a bureaucratically organized system not to perform well are such conditions as (1) inadequate lines of communication within and between different levels of authority in the organization, (2) the failure to acquire fully qualified personnel for any and all positions within the structure, and (3) work procedures which are not properly specified for each task to be performed.

While we may not be in favor of increasing regulation of human behavior it is nevertheless a fact of life in growing

[11]See Max Weber, "Bureaucracy," in H. H. Gerth and C. Wright Mills (eds.), *From Max Weber: Essays in Sociology* (New York: Oxford University Press, Galaxy Book edition, 1958), pp. 196–244.

[12]*Ibid.*; also, cf., Scott Greer, *Social Organization* (New York: Random House, 1955).

[13]Weber, "Bureaucracy."

[14]Cf. Joyce O. Hertzler, *Society in Action*, pp. 182–88, and Arthur L. Stinchcombe, "Formal Organizations," in Neil J. Smelser (ed), *Sociology: An Introduction* (New York: John Wiley and Sons, 1967), pp. 154–202.

societies. An example of the increase in bureaucratization that is common in America today is to be found in Dean Harper's analysis of school systems.[15] Harper sees the following as factors in the bureaucratization of American school systems:

1. *increase in size*—meaning more students, teachers and other personnel, more building, etc.
2. *changing functions*—meaning the transition from simply teaching a few subjects to teaching many, plus the addition of special programs for the handicapped, economically disadvantaged, vocational training, etc.
3. *general bureaucratization of society*—which refers to the fact that our hospitals, factories, and other nonpublic and public facilities have become bureaucratic and thus set a pattern for others.
4. *imposition of bureaucratic attributes from without*—such as the effects of state and federal laws requiring the keeping of certain records, the provision of additional services, etc.[16]

We pay attention to the concept of bureaucracy because modern complex societies are characterized by their bureaucratized institutional facilities. As time goes on and more nations develop into urbanized, industrially based social systems the nonbureaucratic forms of institutional patterns become increaseingly rare.

Since institutions, whether in bureaucratic guise or not, have the primary function of regulating behavior, which means ensuring conformity, it is reasonable to expect that any given institution will be somewhat resistant to change. Bureaucratic institutions should, theoretically, be more open to change than nonbureaucratic ones since they are oriented toward the achievement of efficiency and any alteration in the normal environmental conditions of a society will produce some need for change. However, some writers have noted a tendency toward conservatism in bureaucratic structures which appears to be characteristic of their actual nature.[17]

[15]Dean Harper, "Growth of Bureaucracy in School Systems," *American Journal of Economics and Sociology*, XXIV (July, 1965), 261–71.

[16]*Ibid.*, pp. 262–64.

[17]See, e.g., S. C. Dube, "Bureaucracy and Nation Building in Transitional Societies," *International Social Science Journal*, XVI (1964), 229–36. In this article Dube notes the highly developed class bias and the antipathy toward nationalism in the bureaucracies of developing countries during their colonial phase. Also cf., Peter M. Blau, *Bureaucracy in Modern Society* (New York: Random House, 1956), pp. 85–100.

Social systems develop problems when their institutional network as a whole, or in part, fails to meet the needs of either individual members of a system or the system at large. In the next chapter some of those problems and contributing factors other than the institutional structure will be discussed.

Problem Questions

1. List three different group activities in which you have engaged during the past week. What roles did you play in each? What roles constituted the role-sets of each? Were all these activities related to the same institutional area or to different ones?

2. Select an institutional facility (school, business firm, factory, governmental agency, etc) in the local community and describe it in detail according to the institutional characteristics designated by Malinowski. What part does it play in meeting the goals of the institution it represents? Is the facility operated on bureaucratic principles? If not, how does it operate? Does the facility appear to be flexible enough to allow for changing conditions? If not, what do you think it needs to acquire for greater flexibility?

Selected Adjunct Readings

PART I—ANNOTATED

LOPATA, HELEN Z. "The Life Cycle of the Social Role of Housewife," *Sociology and Social Research*, LI (October, 1966), 5–22.

Dr. Lopata presents us simultaneously with an empirical and theoretical study of the role concept. By concentrating on the major role of housewife she focuses attention on what is meant by role expectations in terms of shifts that occur in those expectations over time. She not only describes the processes by which the role is attained but also its relation to other roles and the changes which must be made at different points in the family life cycle. No better illustration of the interrelationship between role and institution is available at this time.

MEADOWS, PAUL. "The Rhetoric of Institutional Theory," *Sociological Quarterly*, VIII (April, 1967), 207–14.

In this excellent analytical article Prof. Meadows describes the structure of institutions in terms of behavior. He discusses the important place of the role concept and notes that each institution has its own distinctive role patterns. He also notes the place of values and the process of communication in the development of institutional forms. He points out the importance of symbols in human life and discusses the effects of technology on human behavior. This is a concise and worthwhile discussion of the entire nature of institutions and is indicative of the centrality of the concept to sociological theory.

SMITH, DAVID. "The Importance of Formal Voluntary Organization for Society," *Sociology and Social Research*, L (July, 1966), 483–94.

Essentially this is a discussion of the part played by formally organized groups in modern industrial society. The author states that such organized groups make four kinds of contributions (1) provision of goal attainment functions, (2) integration of society at various levels including international, (3) socialization and education, and (4) the provision of latent adaptive facilities. This is a good article for the student interested in the relationship between institutions and their facilities.

WHYTE, WILLIAM FOOTE. "Models for Building and Changing Organization," *Human Organization*, XXVI (Spring, 1967), 22–30.

The purpose of this paper is to suggest that all organized activity is conducted on the basis of models of organization which are implicit, at least, in the behavior of persons who make key organizational decisions. Thus, if organizational decision makers act inappropriately the efficiency of the organizational structure is reduced. Professor Whyte believes more flexibility and imagination can counter such problems. He provides a system for more effective organization building which includes such components as (1) examination of basic activities, (2) relation of personnel to one another, and (3) becoming aware of existing types of organizational models and how they function.

PART II—ADDITIONAL READINGS

ANGRIST, SHIRLEY S. "Study of Sex Roles," *Journal of Social Issues*, XXV (January, 1969), 215–32.

BITTNER, EGAN. "The Concept of Organization," *Social Research*, XXXII (August, 1965), 239–55.

CARNEIRO, ROBERT. "On the Relationships Between Size of Population and the Complexity of Social Organization," *Southwestern Journal of Anthropology*, XXIII (Autumn, 1967), 234–43.

EVAN, WILLIAM. "Organization Lag," *Human Organization*, XXV (Spring, 1966), 51–53.

FRITSCHLER, L. "Bureaucracy and Democracy, The Unanswered Question," *Public Administration Review*, XXVI (March, 1966), 69–74.

HALL, RICHARD H. "Professionalization and Bureaucratization," *American Sociological Review*, XXXII (February, 1968), 92–104.

MOWRER, ERNEST R. "Differentiation of Husband and Wife Roles," *Journal of Marriage and the Family*, XXXI (August, 1969), 534–40.

PALISI, B. J. "Critical Analysis of the Voluntary Association Concept," *Sociology and Social Research*, LII (July, 1968), 397–405.

SCHMITT, RAYMOND L. "Major Role Change and Self Change," *Sociological Quarterly*, VII (Summer, 1966), 311–22.

ZURCHER, LOUIS A. "Sailor Aboard Ship: A Study of Role Behavior in a Total Institution," *Social Forces*, XLIII (March, 1965), 389–400.

8

THE BACKGROUND OF
SOCIAL CHALLENGE

social deviancy and
social disorganization

To a large extent the general public holds the mistaken belief that people whose behavior is deemed undesirable are either suffering from some sort of illness or are innately (often meaning of racial origin) evil. Such ideas, however, ignore certain facts.

... career criminals are not "sick" people. They do not exhibit glaring emotional disturbances. Their behavior can be explained, if at all, in cultural and social terms.*

*Russell R. Dynes et al., *Social Problems: Dissensus and Deviation in an Industrial Society* (New York: Oxford University Press, 1964), p. 541.

Social Deviancy

IF A SOCIETY IS TO BE PROPERLY CHARACTERIZED it is necessary to know not just which kinds of behavior are culturally defined as acceptable and desirable but also to know which kinds of behavior are defined as undesirable and against the best interests of the society. Linton's list of culture elements concentrates on acceptable behavior and is, therefore, incomplete.[1] We can, however, add to his list the term *deviancies* to refer to behavior which does not meet established norms. It is important to have such a social category because not all societies view the same acts as unacceptable behavior.[2] For example, it is typical of Western (that is, European-American) morals and legal systems that the taking of fruit from an orchard is a punishable act, if the permission of the owner has not been sought and obtained. In Aztec society, however, a hungry person had the right to take fruit from another's tree so long as he took it solely for his own needs. On the other hand, it is not unusual to see an intoxicated soldier on our streets and usually even the Military Police will do no more than take him somewhere he can safely sleep off the effects of the alcohol. Among the Aztecs, though, if a young warrior happened to be found publicly drunk he would have been considered to have committed a heinous offense for which he might be punished by being put to death![3]

The basic conception of normative behavior implies that people are expected to hew reasonably well to expected standards, so that *any kind of noticeable variation* could be considered deviant. Consequently deviancy may be said to exist whenever (1) people neglect to do something expected of them, (2) they do too

[1] See p. 48 above.
[2] Cf., W. G. Sumner, *Folkways* (Boston: Ginn and Company, 1940).
[3] See George C. Vaillant, *The Aztecs of Mexico* (Harmondsworth, England: Penguin Books, 1951), p. 126.

84

much or too little, (3) they do something improperly or, (4) their behavior consists of acts completely different than expected of persons enacting their roles. Negative sanctions may be brought to bear in any of the preceding cases but will not if those observing the act do not believe norms have been broken. It should be understood that no matter how reprehensible an act might appear to a given observer it is not deviant unless the cultural values of the system being observed so deem it to be. In general, it can be assumed that acts and attitudes which are at variance with the basic values of a society will be considered deviant *if brought to public attention.* Thus behavior which is hidden from public scrutiny will not be condemned since conformity with societal norms and values is primarily of concern only when people can be observed by others. If we take the case of homosexuality as an example it can be noted that homosexuals who conduct their love affairs just at the minimal level of holding hands are open to condemnation when it is done in public. On the other hand, those homosexuals whose public behavior is not indicative of their sexual inclinations can avoid negative sanctions.

It should be understood that not all passive (female role) male homosexuals have or display publicly any feminine behavior traits whatsoever. Some, in fact, appear in public as athletic "real he-man" types of individuals. In the case of active (male role) female homosexuals, likewise, there is no reason to assume any public display of usual masculine mannerisms of any kind.

What we are getting at here is that the context in which an act takes place has an important bearing on whether it is to be considered a deviancy or not. To put things into proper perspective the social analyst must ask at least the following five questions:

1. *What was done?* The specific nature of the act.
2. *How was it done?* The particular methods used.
3. *Who is involved?* Both the persons committing the act and others affected by their actions.
4. *Where did it take place?* Location in terms of both spatial coordinates and cultural features.
5. *When did it take place?* Date, time of day, season of the year, etc.

As an example, if we hear that one man has struck another we would certainly want to know if it was merely a friendly, though husky, pat on the back or a solid right to the jaw. If the latter, we would certainly be interested to learn if bare or gloved hands were used, or if the hitter had a roll of coins concealed in his fist. Suppose now that boxing gloves were worn, that *might* indicate a sporting event; unless the man punched was over 60 years of age and frail, and his assaulter only 20 years old and very muscular. But, perhaps the whole thing took place in a home, then it might be the case of a grandfather doing too well at teaching his grandson the rudiments of boxing. Still, there's the element of time to consider. Would the old boy be teaching his grandson boxing at three o'clock in the morning? As we can see, any variation in the context is a very important consideration when making a decision about whether an act is deviant or not.

At this point in our discussion it is very important to comprehend that not all acts of deviancy are illegal or immoral. An act is deviant not because it is against the law or at variance with some moral code but rather because it goes against accepted standards of behavior *no matter what the nature of the activities engaged in happen to be*. Edward Manet and Claude Monet, for example, were deviants in the art world at the time they made their great works.[4] Likewise Galileo and Einstein were deviants in science[5] as was Semmelweiss in the field of medicine.[6] Some of these people were considered immoral by their contemporaries but the major reason for the view of them as deviants came about because of their challenge to conservative aspects of the institutional structure of their societies. Manet and other impressionists, for instance, were seen as upstarts who dared to vary from the preferred artistic procedures and values of the French Academy.[7] Thus deviance, like beauty, is largely in the eye of the beholder. A deviancy is an act or belief which some people—sometimes entire societies but often persons representing only part of their institutional network—consider to be improper or injurious to themselves or to the whole social system.

[4]See George E. Slocombe, *Rebels of Art: Manet to Matisse* (Port Washington, N.Y.: Kennikat Press, 1969), pp. 36–42 and 62–64.
[5]See George Gamow, *Biography of Physics* (New York: Harper, 1961) pp. 46–50, and Leopold Infeld, *Albert Einstein: His Work and Its Influence on Our World* (New York: Scribners, 1950), pp. 1–7 and 112–30.
[6]See C. D. Haagensen and W. E. B. Lloyd, *A Hundred Years of Medicine* (New York: Sheridan House, 1943), pp. 291–92.
[7]Slocombe, *Rebels of Art, op. cit*

Deviancy can be the unique act of a particular individual or it may represent a clash of interests between entire social systems (there is no really good reason why war should not be considered deviant behavior when we look at societies which have considerably advanced beyond the primitive level). In the latter case, however, often only the members of one of the clashing systems will consider an act to be deviant. During World War II the Germans did not consider Nazi aggression against Norway, for example, to be deviant from their standpoint. There is also the instance of part of a social system acting in opposition to the larger system as a result of different values being held. If we look at this case in terms of goals to be attained and means of obtaining them, with respect to the legitimacy of each *as defined by the larger or more inclusive system* we end up with the following:

GOALS (as defined by the larger system)

MEANS (As defined by the larger system)	LEGITIMATE (Things to be strived for)	ILLEGITIMATE (Things not to be strived for)
LEGITIMATE (Acceptable ways)	not deviance	deviancy
ILLEGITIMATE (Unacceptable ways)	deviancy	deviancy

We can see then that of the four possibilities described only one can result in *nondeviant* behavior, *from the standpoint of the larger social system*! Thus, in three cases out of four, negative sanctions (some of which might be very severe) could be brought to bear on the deviant persons. The diagram somewhat imperfectly describes Durkheim's contention that social constraints are a major basis for conformity we find in most social systems.[8]

Social Disorganization

Deviancy and social disorganization are not equivalent terms. Deviancy may be the result of acts by individuals or groups but the concept of social disorganization applies to the operation of

[8]See Durkheim, *Sociology and Philosophy*, pp. 41–45.

social systems rather than to the actions of individual persons. *Social disorganization* may be defined as that state or condition such that a social system or a part of one is experiencing a disruptive effect on its basic activities and relationships among its members. That is to say the norms which guide those activities and relationships are not generally being held to. As a consequence, then, the system must either bring pressure to bear in an attempt to reestablish a satisfactory degree of conformity; or to maintain its existence without undue internal conflict, its values and norms must be altered so as to satisfy at least some of the requirements of the population which led to a disruptive condition.[9]

An example of social disorganization in American society is to be found in the prevailing conditions in minority group relations. While we give lip service to the *ideal norm* that all persons, regardless of racial or ethnic background, are equal under the law and have equal rights to avail themselves of any or all opportunities to improve their individual or collective lots the fact is that both open and subtle discrimination do and have played a large part in the real patterns of treatment of minorities.[10] As a consequence of the discriminatory practices applied to minorities the United States has had to follow the route of creating special laws to ensure the rights of minorities so that the system can be said to be bringing pressure toward conformity with the norms!

In fact, of course, the actual degree of conformance varies with the extent to which so-called civil rights laws are enforced by the appropriate officials, as well as to the extent to which any attitudes happen to be changed in the direction of conformity as a result of proper enforcement of the law. It is interesting to note, by the way, that most "law and order" types of political candidates rarely if ever have civil rights laws in mind during their campaigns for public office.

Social deviancy can, if sufficiently extensive, serve as an index of social disorganization but, on the other hand, social disorganization may be productive of deviancy. There is, in short, an interdependency or transactional effect between social deviancy and social disorganization.

---

[9]See Russell R. Dynes, Alfred C. Clark, Simon Dinitz and Iwao Ishino, *Social Problems: Dissensus and Deviation in an Industrial Society* (New York: Oxford University Press, 1964), pp. 4–10.
[10]*Ibid.* pp. 351–58.

Drug addiction, delinquency, and functional mental illness are all examples of deviant behavior which can be related to pervasive conditions of social disorganization. What are the rates of these three deviancies when the United States is not engaged in an unpopular war, or if there is no massive inflation, or if unemployment is at a considerably lower level?

Compare, for example, arrests for violations of the illegal narcotics laws at the beginning and at the end of the 1960s. There were 31,000 such arrests in 1960 and 153,000 in 1969, a 491.9 percent increase over the decade![11]

These interrelationships indicate inadequacies in our institutions which must be corrected if we are to cope effectively with the aforementioned problems or any others.

It is also true, however, that deviancy and disorganization can be produced by natural occurrences, such as floods or earthquakes, or may result from various kinds of sociocultural change. Certainly the advent of the steam engine and the concept of parts interchangeability resulting in the Industrial Revolution produced massive imbalances and disruptions in American society in the last century—but so did the expansion of the nation westward. That expansion, among other things, led to a great deal of lawlessness in the western territories. Social disorganization and social deviancy then are continuously occurring to some degree or another in all societies. They become major social problems, however, only when they occur at a massive level or in those instances when the public becomes particularly concerned about their effects.

Problem Questions

1. Describe two different kinds of deviancy in which you have been involved. Specify the characteristics of the situation (What, How, Who, Where, When) which led to the definition of your behavior as deviant.

[11]See U.S. Bureau of the Census, *Statistical Abstract of the United States: 1971* (92nd. ed.) (Washington, D.C.: U.S. Government Printing Office, 1971), p. 146, Tables 231, 232.

If illegal acts were not conducted which social system was responsible for defining you as a deviant?

2. Search the Public Affairs and Social Science Indexes (ten years back) and make up an annotated bibliography on drug addiction. Try to locate any sources which discuss the relationship of the tobacco industry to addiction. If you cannot locate any, what reasons can you give for their scarcity?

Selected Adjunct Readings

PART I—ANNOTATED

BRIAR, SCOTT and IRVING PILIRAN. "Delinquency, Situational Inducements, and Commitment to Conformity," *Social Problems*, XIII (Summer, 1965), 35–45.

The authors of this important article see delinquency primarily as acts of opportunity, motivated by the particulars of a situation. Younger boys, for example, strongly influenced by their associates will commit acts that they would be reluctant to perform at a later age when they become eligible for the job market. They want to conform with their peers, so if the latter are delinquent they become so. When employment opportunities become a salient influence, however, many youngsters drop their delinquent traits and consequently induce others to do likewise. This article is a good illustration of the effects of subsystem culture on behavior.

MARWELL, GERALD. "Adolescent Powerlessness and Delinquent Behavior," *Social Problems*, XIV (Summer, 1966), 35–47.

This is a good follow-up to the preceding article. Marwell notes, for example, that adolescents are recruited into delinquent roles by other adolescents. He implies that, in part, the recruitment is not difficult because the youngster feels a need to react against adults and to reveal his own maturity via what *he considers* to be a proper masculine role for example. He has been controlled by others all his life and he now wants some power of his own. The delinquent peer group may provide him with ways of demonstrating his power and his sex status. This article also contains some ideas for treatment and prevention of delinquency.

QUINNEY, RICHARD. "Suicide, Homicide and Economic Development," *Social Forces*, XLIII (March, 1965), 401–6.

This author conducted an analysis of the suicide and homicide rates of forty-eight countries in order to determine if there was some relationship between the rates and levels of economic development. The indications of this study are that if economic development is measured by urbanization and industrialization there is a pattern of relationship to suicide and homicide rates that is evident. However, it was also noted that while suicide varies directly with development, homicide takes an inverse relationship, thus indicating that the two acts represent different kinds of social phenomena.

PART II ADDITIONAL READINGS

ALEXANDER, C. NORMAN, JR. "Alcohol and Adolescent Rebellion," *Social Forces*, XLV (June, 1967), 542–50.

BATES, WILLIAM M. "Narcotics, Negroes and the South," *Social Forces*, XLV (September, 1966), 61–67.

BOGGS, SARA L. "Urban Crime Patterns," *American Sociological Review*, XXX (December, 1965), 899–908.

ERIKSON, KAI T. "Notes on the Sociology of Deviance," *Social Problems*, IX (Summer, 1962), 307–14.

FLEISHER, BELTON M. "The Effect of Income on Delinquency," *American Economic Review*, LVI (March, 1966), 118–37.

GANNON, THOMAS M. "Religious Control and Delinquent Behavior," *Sociology and Social Research*, LI (July, 1967), 418–31.

MARSHALL, TONY F. and MASON ALAN. "Framework for the Analysis of Juvenile Delinquency Causation," *British Journal of Sociology*, XIX (June, 1968), 130–42.

MORRIS, R. B. "Female Delinquency and Relational Problems," *Social Forces*, XLIII (October, 1964), 82–89.

REISS, IRA L. "Premarital Sex as Deviant Behavior: An Application of Current Approaches to Deviance," *American Sociological Review*, XXXV (February, 1970), 78–87.

VAZ, EDMUND W. "Juvenile Gang Delinquency in Paris," *Social Problems*, X (Summer, 1962), 23–31.

VOSS, HARWIN. "Socio-Economic Status and Reported Delinquent Behavior," *Social Problems*, XIII (Autumn, 1966), 314–24.

WECHSLER, HENRY. "Community Growth, Depressive Disorders, and Suicide," *American Journal of Sociology*, LXVII (July, 1961), 9–16.

9

PATTERNS OF
SOCIAL DYNAMICS

social change and
social trends

No society, no matter how isolated, nor how traditional, nor how well controlled remains exactly the same over time. New people are born, old ones die, natural conditions such as the weather can cause modifications of behavior, discoveries of a technical or scientific nature are made by chance, and so on it goes. For human societies, change is a constant.

The aboriginal attitude toward peace as a value was being challenged by militarism. Warfare, from being purely defensive ... was becoming offensive ... a number of individuals were withdrawn from activities directly related to subsistence ...*

*Paul H. Ezell, "The Hispanic Acculturation of the Gila River Pimas," *American Anthropolgist*, LXIII, Memoir 90, (October, 1961), 136. Reproduced by permission of the American Anthropological Association.

Social Change

THERE IS NO SUCH THING as a completely static social system. Modifications in the *structure, organization,* and *values* of a system are always ongoing although the rates of change will be much slower in some systems than in others. Our first consideration in the study of social change, however, is not the matter of relative rapidity but rather the question of what the primary sources of change happen to be. We can denote four general primary sources of social change in relation to the total environment.[1]

1. *Natural-physical:* comprised of such occurrences as modifications in climate (e g., a greater or lesser amount of rainfall than normal), geological disturbances (e.g., earthquake or landslides), or increased solar radiation.
2. *Natural-biological:* which consists of conditions such as epidemics or epizootics, sizeably increasing or decreasing population (of humans, plants, or animals), or, perhaps, genetic alterations in some organisms.
3. *Social-psychological:* which refers to such phenomena as rumors, miraculous visions or revelations, or attitude and opinion shocks (e.g., the "credibility gap").
4. *Cultural-technological:* which has to do with technical and scientific innovations of any kind from the percussion formed stone projectile point to the laser beam.

The first two are largely responsible for the latter two but advancing technology has allowed man to produce alterations in the natural environment also, for example, air and water pollution. Largely however, natural-physical and natural-biological conditions occur without or despite man's actions. Social-psychological and cultural-technological conditions are the result of

---

[1] See pp. 12–13 above.

human beliefs and values, discoveries and inventions. Beliefs and values set attitude patterns so that people become more or less desirous of and, hence, more or less receptive to change of any kind. Discoveries and inventions may produce alterations in beliefs and values *if* the new ideas and objects are sufficiently attractive to overcome whatever traditional attitudes exist that operate against change in a system.

Discoveries, which are often accidental, and *inventions,* which are generally deliberate, may be either material or non-material in essence. That is, they may be concrete objects or abstract ideas or sometimes a combination of both (as in the case of Whitney's development of the interchangeables part system for mass manufacture).

Some innovations are developed with a social system and some come from outside. The former case we refer to as "independent discovery" or "invention" and the latter as "cultural diffusion."

The reader should note that although we generally credit inventors with creativity and high intelligence, important *discoveries* are sometimes made by people who are just sufficiently alert and intelligent enough to comprehend that they have indeed found something new—for example, William Roentgen and the discovery, *not invention,* of X-rays. Invention implies the development of new uses for known objects or the combining of two or more known objects (material or not, ideas are objects too) into something new. Eli Whitney's idea of constructing military rifles out of parts which could be interchanged, because each part is made within certain limits of dimension and material, was the essential invention before mass production could become a reality. Generally speaking, the process of inventing does imply creativity and intelligence but some inventors have lacked one or both qualities to a high degree and only sheer luck has enabled them to accomplish what they set out to do. An example of this is Goodyear and the development of the process for vulcanizing rubber in order to make it a genuinely useful material.[2]

That diffusion is an important process is unquestioned but there have been times when excessive importance was placed on

[2]See Mitchell Wilson, *American Science and Invention* (New York: Bonanza Books, 1960), pp. 124–29.

particular societies as *the* centers from which all good ideas sprang. Diffusionist "schools" such as the *heliocentric* (meaning sun-centered) consisted of people who believed that all of mankind's major developments came from one of a few special centers. In the case of the *heliocentrists* the center of all civilization and culture was ancient Egypt. The heliocentrist's believed that since pyramids were found in ancient Egypt and also in Central America and that since the Egyptian pyramids came earlier in time the Egyptian's must have crossed the ocean and taught the American Indians how to build pyramids. There are two fallacies inherent in that belief, however, and one of them is the question of structure versus function. For example, the pyramids of Egypt bear a superficial resemblance to those of Mexico; that is, they have the same basic geometric form. The Egyptian pyramids, however, are tombs and true buildings, while the Mexican ones are actually stone covered hills upon which religious temples were erected in order to be closer to the heavens. Thus the two kinds of pyramids are *functionally different*. The second fallacy lies in the time factor. The latest known Egyptian pyramids in terms of time of construction are about two thousand years older than the oldest known pyramids in Central America or Mexico. Thus, even if the Egyptians had sailed to the Americas why did it take two thousand years for the American natives to learn how to construct a pyramid, let alone one quite different in function from the older type.[3]

Thor Heyerdahl has recently tried to show the possibility of a trans-Atlantic crossing by sailing his papyrus raft Ra (named after the Egyptian sun god) from Africa to the Americas but he proved *no definite culture contact* by the efforts of himself and his courageous crew.[4]

Nevertheless, diffusion is a fact and some societies have made major contributions to all mankind although no society is or likely ever will be, responsible for all or even a majority of human achievements. If we take food, especially vegetable

[3]For an interesting discussion of the structure versus function question see G. Elliott Smith, Bronislaw Malinowski, Herbert J. Spinden, and Alexander Goldenweiser, *Culture: The Diffusion Controversy* (New York: W. W. Norton and Company, 1927), especially Malinowski's remarks, pp. 26–46.

[4]See Thor Heyerdahl, *The Ra Expeditions*, translated by Patricia Compton (Garden City, N.Y.: Doubleday Inc., 1971).

foods, as an example Asia and Central and South America loom very large in the overall picture of the development of food plants. So far as is known the peoples of Asia are probably responsible for giving to the world in cultivated form: apples, peaches, pears, plums, cherries, bananas, oranges, lemons, limes, mangoes, carrots, lettuce, muskmelons (including the canta- loupe and honeydew varieties), onions, soybeans, and rice. Now this is an impressive list of important food crops but when compared with the contributions of the native Americans it may seem not so great. For the American Indian has given the world the following plant foods: white("Irish") potatoes, sweet potatoes, lima beans, kidney beans (many types), string (green or snap) beans, pineapples, cranberries, avocadoes, corn (maize- many kinds), tomatoes, green peppers, red peppers, the enormous squash family (including pumpkins), vanilla, cocoa, papaya, and so-called "wild" rice, the gourmet's delight.

When we realize that Marco Polo first brought spaghetti (and other forms of compressed wheat products) into Italy from China in the thirteenth century and the Italians had to wait several centuries more for Columbus to go to the new world so the Spanish could bring back the basic ingredients for the sauce that has made spaghetti a world wide favorite we can see the effects diffusion (from Asia and the Americas to Europe) as well as invention (spaghetti sauce of the basic tomato type— the unique Italian contribution).

In the case of mechanical or other "hard" inventions the apparent superiority of some societies is only a surface mani- festation. If we take the example of the airplane there is no question of the important position of the United States in avia- tion history. Yet some American inventors such as Langley were actually in the realm of failures.[5] Furthermore, when it came to manufacture and the use of aircraft the United States was far behind other nations within a dozen years of the Wright brother's successful flight in 1903. For during World War I the United States, for all its vaunted industrial-technical prowess, did not develop nor successfully build any major kind of military combat aircraft![6] In fact American industry did such a poor job

---

[5]See, e.g., Martin Caidin, *Air-Force* (New York: Bramhall House 1957), p. 2.
[6]*Ibid.*

of building the British De Havilland military plane that most of those shipped overseas had to be torn down and rebuilt![7]

Going beyond that period we find much experimentation with aircraft and techniques of using them for peace and war in the 1920s and early 1930s but shortsighted military and civilian officials kept the United States from becoming a first rate air power until after the Japanese attack on Pearl Harbor. We can note, for example, that the major early World War II fighter plane of the U.S. Army Air Force was the Curtiss P-40 which, while very sturdily built, lacked a supercharger for high altitude operations and which, in its initial combat form was greatly underarmed. On the other hand, the U.S. Navy was using as its first line fighter the rugged little Grumman Wildcat which, when equipped with full combat gear, could not attain 300 miles per hour in level flight! In addition, some American Marine fighter units were stuck with the Brewster Buffalo, a plane which lacked just about everything for survival in aerial combat against the superior Japanese aircraft.[8]

This picture of the American air industry and military air forces before World War II demonstrates the place of vested interests, traditionalism, and public apathy in resistance to change.

Innovations of any kind are accepted within a social system only if:

1. the culture is broad and complex enough for there to be some likelihood that it will be possible to comprehend the usefulness of an innovation.
2. prevailing attitudes of large segments of the population are not such that any change will be resisted or rejected.
3. there are no powerful vested interests that can prevent a particular innovation from being made known to most of the system members, or if known, to prevent its adoption.

Sometimes, of course, members of a system will actively promote changes they desire. They may do this by propagandizing the public, through the mass media and by organizing a group of people to actively support change by public demonstrations, or writing letters to officials, etc. Such organized efforts

---

[7]*Ibid.*
[8]See Martin Caidin, *Golden Wings* (New York: Bramhall House, 1959), pp. 94–97 and 120–21.

to promote change are called "social movements." Some movements have massive support initially and some have very little, but all go through the same processes of trying to attain their particular goals. Once some members of a system decide certain needs should be met in certain ways they usually first try out their ideas on the general public. If the ideas seem acceptable to a sizeable minority the initiators of the movement will try to recruit others to their ranks. If successful in that effort they will proceed to do the following:[9]

1. present a continuous barrage of propaganda to promote their ideas.
2. appeal for funds, votes, and whatever else will help the movement gain economic and political strength.
3. take what action seems necessary to achieve initial goals, such as boosting candidates for political office they believe are sympathetic to them or by attempting to disrupt elections or legislative bodies or other official functions they see as detrimental to their cause.
4. if they gain initial goals and acquire the strength to attain the desired changes the movement will either fade into the background or seek new goals to maintain its existence.

Note that not all social movements are directed toward attaining something entirely new. Some movements in fact are past oriented, they want to hold on to traditional ways they believe are being eroded, or sometimes even want to go back to old or discarded modes of action. For example, many fundamentalistic "revival" groups are of this type, as are some political organizations such as the American Independent Party.

By way of contrast the Richard M. Nixon administration, while hardly falling into the "liberal" political category, has produced proposals, such as the idea of a minimum welfare payment, which, being essentially new, should be classified as progressive from the social change standpoint. It might also be noted here that President Nixon did campaign, in part, on a pledge to give eighteen-year-old persons the right to vote, something to which John F. Kennedy had voiced opposition. The point is that the temper of the times and past occurrences have a major effect in

---

[9]Cf. the discussion of ideal-typical stages of social movements in Joyce O. Hertzler, *Society in Action* (New York: Dryden Press, 1954), pp. 369–71.

determining whether an idea or an action represents change we can call "progressive" or "retrogressive." Change directed toward obtaining the new (that is, what is considered new in a particular system) we can refer to as *progressive change* and movements with this orientation can be called *progressive movements*. Change directed toward previously existing ideas we can refer to as *retrogressive change* and social movements having such an orientation can be called *retrogressive movements*. The terms "progressive" and "retrogressive" do not connote "good" or "bad" but only the direction of change. Neither can we assume that progressive movements are necessarily *liberal* nor retrogressive ones *conservative* in the political sense. For example, the late President John F. Kennedy's "New Frontier" was largely a rehashing of ideas first propounded by the New Dealers under Franklin D. Roosevelt. Thus, although Kennedy's proposals were "liberal" they were also, to some extent, retrogressive though inclined to aim toward goals not successfully achieved under the Roosevelt administration.

Some Theories About Social Change

Social theorists have come up with a number of ideas about social change.[10] Among these is the view that change can be looked at in terms of the stability or equilibrium of a social system. That is, a social system will be assumed to be static until something disruptive to its present state occurs and sends it into the throes of adaptation to the disturbance. Smelser has discussed equilibrium models of change and noted a range of types.[11] He also indicates that he believes the equilibrium viewpoint may be the best approach to analyzing social change because all variables can be considered.[12] Essentially he sees change occuring in three phases (1) an initial impetus, (2) a recovery attempt stage, and (3) long range effects.[13] If we take as an example

[10]See, e.g., Richard P. Applebaum, *Theories of Social Change* (Chicago: Markham, 1970).

[11]See Neil J. Smelser, "Toward a General Theory of Social Change," Chapter 8, in his *Essays in Sociological Explanation* (Englewood Cliffs, N.J.: Prentice-Hall, 1968), especially pp. 263–65.

[12]*Ibid.*, p. 268.

[13]*Ibid.*, pp. 269–78.

a worsening state of inflation in an economic change situation, as the initial impetus, we might find the government of a country attempting to adjust its economy by placing special tariffs on imports or by imposing wage and price controls, or both. Should such attempts at recovery fail, then more severe conditions, such as a recession, might come into being and force the society to adopt more stringent measures, or perhaps even to change the whole nature of its economic system. There might be a shift from capitalism to socialism (or vice-versa) or industrial production might become more diversified, such as a concentration on items for trade. Equilibrium theorists might well hold that adjustment to such changes is part of the nature of social systems and, therefore, systems which would not adjust would eventually cease to exist.

Another type of change theory looks at imbalances in the relations between technology and social controls. William F. Ogburn formulated the "cultural lag" hypothesis in which he perceived social problems arising from the fact that material culture often develops beyond the ability of a society to adapt to it.[14] For example, the major world powers are now grasping for ways to control the spread and use of nuclear weapons. On the other hand, there is "achievement lag," which implies the existence of ideas for technological advances but coming at a point in time when there is a definite lack of the resources and techniques necessary in order to obtain the desired goals.[15] For example, medical men would like to find a cure for the common cold, an idea they have been pursuing for a long time, but apparently the present state of medical technology is not sufficient for the task. We might also consider the hypothesis of "organizational lag" wherein technology and ideas for its use and control get ahead of the ability to coordinate efficiently all the procedures necessary to obtain a given goal. For example, the use of computers in some bureaucratic organizations has sometimes led to less, rather than more, efficiency because the organization was not initially set up to operate on a basis other than sheer manpower, and the installation of computers requires both a shift in skills and a change in management patterns.

[14]William F. Ogburn, *Social Change* (New York: B. W. Heubsch, 1922).
[15]See Howard P. Odum, *Understanding Society* (New York: Macmillan, 1947).

Probably a combination of *equilibrium* and *lag* theories would be very fruitful in analyzing social change. The basic problem, however, lies in the specification of the variables to be included and the nature of the systems to be examined. In regard to the latter, Howard Becker once described four types of societies in terms, among other characteristics, of their relationships to change. Those societies are[16]

1. *Proverbial*—marked by almost total isolation from outsiders, thus holding diffusion and acculturation to a minimum. Such societies are highly traditionalistic and, hence, any innovations are viewed with great suspicion.
2. *Prescriptive*—not quite so isolated as the first but a great deal of formal regulation of intercourse with outsiders prevents diffusion of culture traits which the leaders view as undesirable. Internal innovation is similarly restricted.
3. *Principal*—generally oriented toward change in both its basic values and practices and the more formal aspects of its culture. In this kind of society there is some traditionalism but it serves as a brake to change rather than standing as a wall to prevent its occurrence.
4. *Pronormless*—totally involved in change to the extent that one day's norms may be outdated the next day. Change takes place so rapidly here that the society is very much disorganized, perhaps tending toward disintegration.

Becker was trying to deal with the problem of change in whole societies but what about small social systems? They too undergo change. What factors then are common to all systems regardless of their complexity? We can probably not too well utilize less than the following three variables (and still be relatively comprehensive in our study of social change)

1. *Core values*—concerns the primary goal orientations of the system
2. *Cultural components*—the extent of material and nonmaterial cultural elements available within the system
3. *Situational concerns*—needs of the system during the period of time in its history of interest to the social analyst

If the preceding three factors are known it is no great prob-

[16]See Howard Becker, "Current Sacred-Secular Theory and Its Development," in Howard Becker and Alvin Boskoff, eds., *Modern Sociological Theory: In Continuity and Change* (New York: Dryden Press, 1957, pp. 133–85.

lem to understand the changes that take place in a system or the rates of change that occur. However, it should be noted that these factors can be considerably difficult to cope with in regard to the ease of obtaining sufficient information or the comprehension of multifaceted relationships.

We can also look at change in respect to different levels of social complexity within a given large scale system. If, for example, we take an entire nation as a starting point how can we relate changes on the national level to their effects on particular personalities?[17] One way of handling this problem might be to look at the diffusion of change through a series of intervening systems such as the following:

SYSTEM LEVELS

Simple ◄———————————————————————► Complex

Person	Peer Group	Community	Region	Nation
		CULTURE LEVELS		
Personality	Subcultural Patterns			Mass Culture
		CHANGE PROCESSES		

◄——————————— inward from mass culture

outward from individual personality ——————————————►

This diagram is a very simplified illustration of the fact that social change is an ongoing process of interaction between persons and groups, and between groups of different sizes.

There is also the question of spheres of change that affect all kinds of social systems. If we were examining large scale systems alone we might well look at particular institutions, but institutions would not be a good starting point for the examination of change in rather small scale systems. There are three areas of change, though, which are important for systems of any size or degree of complexity. They are:

1. *Population change*—alterations in the size, distribution, rates of growth, or compositions of a population can have profound effects on a social system. E.g., high survival rates for male and low survival rates for female babies could, within a generation or so, produce a society strongly leaning toward polyandry— the marriage of a woman to several men.

[17]See pp. 43–44 above.

2. *Ideological change*—here we are concerned with changes in any realm of thought (not just politics as some people define ideology). We might, for example, look at the world of art. In recent years some profound changes have taken place in what artists, critics, and the public consider to be acceptable forms of art and modes of presenting art works for appreciation.[18] Large pieces of rusty metal crudely welded into an almost formless mass may now be the object of considerable admiration—and they may be presented not within the confines of a museum but placed out on the street or in the lobby of a business building. Such alterations in thinking can eventually lead to a major restructuring of social relationships in terms of roles and statuses.

3. *Technological change*—in this case we are dealing with advancements in material culture which includes both artifacts and the techniques for their production and application. Major technological innovations such as the automobile and television have had a highly visible impact on Western society, and especially American society. Even more profound alterations have been created by three related developments; namely, improved mass production techniques (for example, automation), the unionization of industrial workers, and increased efficiency in all phases of industrial development resulting from the use of computers and other high speed calculating and record-keeping devices.

It can be noted in respect to these three areas of change that, for example, before the United States could begin to become a major industrial power the following changes had to occur:

1. Efficient machines and mass production techniques had to be developed.
2. Sufficient manpower with necessary skills had to become available.
3. Enough capital had to be invested and resources made available for industrial purposes.
4. A shift had to occur in thinking from basically agricultural values to those more characteristic of a complex system of manufactures.

[18]For an example of change oriented research in the realm of the arts see Richard B. Vexler, *Jazz Music in the United States—The Institutionalization of an Art Structure* (unpublished M.A. thesis, University of Pittsburgh, 1961). Also see his "Jazz as Art: The Critics and the Public," *New York Folklore Quarterly* (March, 1968).

There are three points in this discussion of social change which should be kept in mind. First, change occurs relative to the culture of a given system at any point in time. Second, change is multi-directional in that it can be oriented toward new or old ideas. Third, change can arise from conditions occuring within a system or as a result of contact with other systems.

Social Trends

On the basis of what has been discussed in the section on social change what can we say about the future of any social system? Lets look at the United States as an example of a large scale highly dynamic social system. As of 1970 there were somewhat over two hundred million people in the United States. Around ninety million of these were under 25 years of age and about twenty million were over 60 years of age. This means that the majority of the population was either very young or very old and both of these age categories are tending to increase at the expense of the middle-age range. It is quite possible that should the present demographic situation continue to the end of the century then out of a population of approximately 285 million people[19] in the year A.D. 2000 there may be as many as 130 million persons under age 25 and thirty million over age 60. What do these figures mean in regard to societal transformations? For one thing we may logically surmise that thirty years from now there will be a need for a greatly expanded educational system and that training will become more highly technical and complicated for almost any kind of work. Therefore, many young persons would not begin work until their middle or late twenties and thus would be dependents of their families or the government for some time. This would necessitate a high increase in economic productivity in order to provide for these people during their long period of training and taxes would undoubtedly be on the rise to fund government programs. Furthermore, their late entry into the full time work

[19]This is the present authors projection and may be very far from the actual total figure at that time.

force would mean a shorter period of productivity for most of them.

In addition, with so many older people and a general tendency, even now, for retirement to come comparatively early (age 55 is not an uncommon retirement age in some industries), the society will have to provide much more in the way of leisure time activities for older people. Mark, however, that work weeks for everyone are becoming shorter, thus the leisure and recreation industries may have to show considerable expansion by A.D. 2000. In terms of technology these trends would seem to indicate that innovations in the fields of education and leisure will have high interest priority in the early twenty-first century.[20] Thus, the teacher and the performer of that time could reach a height of status unknown since the time of the classical Greek civilizations. As for the accuracy of these predictions we will just have to wait to find out.

Problem Questions

1. Select a topic on social change such as student movements on campus, conservation efforts, effects of television, etc. Collect a bibliography of at least 10 references from professional literature and see if you can answer the following questions.
   a. What public attitudes support or resist change?
   b. What vested interests are actively opposing change?
   c. What is the nature of the change patterns you are examining? That is, is the change primarily progressive or retrogressive?

2. Think of some change you would like to take place in your home community. What procedures would have to be followed to attain your goal? Which persons in the community would be most likely to be helpful to your cause and which most in opposition? What has happened in other communities that could help in your endeavors to produce change in your own?

[20]We could also logically presume considerable expansion in the area of health care and in the provision of daily care of children whose parents are in school, working, or perhaps on vacation without them.

Selected Adjunct Readings

PART I—ANNOTATED

ALLEN, FRANCIS R. and KENNETH W. BENTZ. "Toward the Measurement of Sociocultural Change," *Social Forces,* XLIII (March, 1965), 522–32.

The authors of this article developed an instrument for measuring social change in America. For the period 1940–1960 they used 32 indicators for the assessment of change in several areas of American society. Their indicators revealed four basic components necessary for change analysis, (1) alterations in standard of living, (2) growth of population, (3) industrial-technological-urban development and, (4) increasing educational levels. These components were seen to be consistent with the general American value system and the authors note, therefore, that their procedures may not work in societies having different basic values.

DUMOND, D. E. "Population Growth and Cultural Change," *Southwestern Journal of Anthropology,* XXI (August, 1965), 302–24.

This very excellent essay notes that human fecundity, a noncultural characteristic, is a basis for increase in human populations. Such increases are seen to lead to the necessity of providing for control of disease, conflict, and subsistence. When subsistence means cannot be expanded the society must either find ways of limiting population growth or accepting a decreased standard of living. The author also states that population increase requires a tightened pattern of social organization and implies, therefore, an extension of all forms of social control. In short, this article implies that increasing population leads to decreasing freedom of action for the individual. Every student should read and ponder this article.

O'CONNELL, JAMES. "The Concept of Modernization," *South Atlantic Quarterly,* LXIV (Autumn, 1965), 549–64.

Mr. O'Connell provides the social analyst with an excellent working definition of modernization. He sees it as the process through which traditional and pre-technological societies pass as they are transformed into secular, rationalistic societies characterized by highly differentiated social structures and machine technology. This author makes the important point that modernization does not come about through a mere accumulation of cultural ingredients but rather that there must develop a state of mind or set of attitudes oriented toward "creative rationality" and away from traditionalism.

PART II—ADDITIONAL READINGS

BROYLES, J. A. "John Birch Society: A Movement of Social Protest of the Radical Right," *Journal of Social Issues*, XIX (April, 1963), 51–62.

CLARK, KENNETH B. "Problems of Power and Social Change: Toward A Relevant Social Psychology," *Journal of Social Issues*, XXI (July, 1965), 4–20.

COLEMAN, A. LEE. "Race Relations and Developmental Change," *Social Forces*, LXVI (March, 1966), 1–8.

DU TOIT, B. M. "Substitution, A Process in Culture Change," *Human Organization*, XXIII (Spring, 1964), 16–73.

EISENSTADT, S. N. "Institutionalization and Change," *American Sociological Review*, XXIX (April, 1964), 235–47.

ETZIONI, AMITAI. "Toward a Theory of Guided Societal Change," *Social Casework*, LXIX (June, 1968), 335–38.

FELLIN, P. "Reappraisal of Changes in American Family Patterns," *Social Casework*, XLV (May, 1964), 263–67.

GUSFIELD, JOSEPH R. "Tradition and Modernity: Misplaced Polarities in the Study of Social Change," *American Journal of Sociology*, LXXII, (January, 1967), 351–62.

LOPREATO, JOSEPH. "Economic Development and Cultural Change: the Role of Emigration," *Human Organization* (Fall, 1962), 182–86.

MAYO, SELZ C. "Social Change, Social Movements, and the Disappearing Sectional South," *Social Forces*, XLII (October, 1964), 1–10.

MOORE, WILBERT E. "Predicting Discontinuities in Social Change," *American Sociological Review*, XXIX (June, 1964), 331–38.

POPE, HALLOWELL and DEAN D. KNUDSEN. "Premarital Sexual Norms, the Family, and Social Change," *Journal of Marriage and the Family*, XXVII (August, 1965), 314–23.

REED, E. R. "Duffusionism and Darwinism," *American Anthropologist*, LXIII (April, 1961), 375–77.

SIMEY, T. S. "The Problem of Social Change, The Docks Industry: A Case Study," *Sociological Review*, IV (December, 1956), 157–66.

APPENDIX

Origins of Modern Sociology

FOR ALL PRACTICAL PURPOSES modern sociology can be said to have had its beginnings in the early nineteenth century when the French metaphilosopher August Comte both coined the word (*sociologie* in French) and defined the term. He thought of sociology as the overall organizing science of mankind which would coordinate and synthesize all knowledge. We modern sociologists have, unfortunately, not yet managed to realize Comte's great aims and have to be content with the study of the various kinds of social interaction that come to our attention.

Early in the nineteenth century only a very few people were writing what can be termed sociological treatises (in the modern sense of sociological works) although a number of persons did write on subjects of considerable interests to sociologists.[1] The first important writer after Comte to deal with sociology as a distinct field was Herbert Spencer. Spencer was originally a railway construction engineer. One day he came to the conclusion that because of his broad reading knowledge and great intellect he should start writing down everything he believed to be the important facets of all human endeavors and accomplishments. Among his early works were articles on political behavior and economics but his first book, published in 1850, was called *Social Statics* and is considered by some to be the first formal work in

---

[1] Notable among these were the biologist Charles Darwin, the political philosopher Henri de Saint-Simon (who was a mentor of Comte), and, of course, the political-economist Karl Marx.

sociology to come into print. In that book Spencer utilized mechanical analogies to provide a theory of societal development. His theory, basically, was that social progress involved a transition from societies characterized by a similarity of constituent units and a simplicity of structure to societies composed of highly differentiated units having a rather complex structure. A decade later Spencer wrote a book of philosophy which he titled *First Principles*. This book supposedly was concerned about physical forces and the characteristics of physical phenomena but, nevertheless, it contained a number of sociological ideas. Eleven years later Spencer's *The Study of Sociology* was published and only three years after that his ponderous *Principles of Sociology*, much less readable than the previous book and containing little, if anything, of new contributions was put into print.

Spencer utilized simultaneously a *mechanistic* (i.e., the idea of interchangeable parts and wholes) and *organismic* (society as a living organism) view of social life. For example, to Spencer society contained a sustaining system (the economy), a distributing system (communications and transportation networks), and a regulative system (government). Thus society is analogous to a living organism. However, there is this difference between a biological organism and human society: the latter has no consciousness of its own as a whole. Instead a society consists of a number of members each having its own consciousness and each of which is therefore replaceable like parts in a machine insofar as the same tasks are accomplished by the replacements.

For all practical purposes, though, Spencer's major contribution was the spurring of interest in the study of society and social relationships by others, including people who did not agree with his conservative political views.[2]

One person who did agree with Spencer's political views was William Graham Sumner. Sumner was one of the founders of both sociology and cultural anthropology in America. He was educated at Oxford and Yale and, at the latter place, he

---

[2]He was a great propounder of the laissez-faire philosophy of politics and economics. He believed the wealthy had attained their status through personal superiority and the poor were so because of innate inferiority. His philosophy of human organic and social evolution is characterized by his famous term "the survival of the fittest" sometimes quite erroneously attributed to Darwin.

received a degree from the school of divinity and was ordained an Episcopal minister. He did not long practice in the church before he became more interested in politics, economics, history, and sociology. He was appointed a professor at Yale and remained there for the rest of his life. Sumner was a follower of Spencer's ideas about economics and politics but unlike Spencer who based almost all his concepts on European societies Sumner collected data on people from all over the world. He presented a selection of this data in semiorganized form in his book *Folkways*.[3] Sumner coined the term *folkways* which refers to customs and habits associated with the routine daily activities common to a given society and he also coined the term and developed the concept of *ethnocentrism* which is defined as the feeling that one's own group is superior to others. Sumner was not a great theorist but he did act as an encyclopedist by bringing together much interesting information about the peoples of the world.[4]

A contemporary and an antagonist of Sumner was the brilliant American philosopher-scientist Lester F. Ward. Ward was a curator of the Smithsonian Institution. His primary specialty was the field of paleobotany—the study of fossil plants—but he studied everything else for which he could find the time![5] He decided at one point in his life that people were at least as interesting as fossil plants, and then proceeded to become a sociologist. Unlike Sumner who felt that social evolution should be allowed to progress by limiting restrictions on individual behavior (i.e., the laissez-faire philosophy) Ward felt that societal progress should be directed by groups of people who possessed varying kinds of expertise. Ward called his system of societal operation "sociocracy." He drew his views of social analysis and social improvement largely from Comte but, interestingly enough, he drew some of his concepts of social evolution from

[3]Published in 1906 but containing material he had collected from the time of the Civil War to the end of the 19th century. The book is in print today in paperbound form and is available from several publishing sources.

[4]Sumner's haphazard collection of all sorts of data led to the development of the Human Relations Area Files at Yale. The *H.R.A.F.* is an important source of ethnological information.

[5]For a fascinating biography of Ward see *Lester F. Ward: The American Aristotle* by Samuel Chugerman (Durham, N.C.: Duke University Press, 1939).

Spencer. Unlike Spencer (and Sumner) Ward believed that the basic difference between humans and lower animals lay in the fact that the latter, acting primarily on instinct, had little or no control over the evolution of their aggregates while humans could and do control parts of their environment, including their societal patterns. Ward summed up his ideas by developing the concept of "telesis" or purposive action. The basic sociological contribution made by Ward is the idea that human beings are not and need not be controlled by their instincts and environment but that, unlike lower animals, they can and do deliberately set future goals to be attained and then pursue them, with the consequence that man alone, of all living creatures, is really capable of changing his ways of life.

In Europe during the latter half of the 19th century a number of people were beginning to concentrate on man's social life as an object of serious study. Perhaps the most important of the European students was a man named Emile Durkheim. Durkheim was a professor at the University of Bordeaux and later at the University of Paris. He studied religion, work and industry, societal evolution, economic history, politics, and deviant behavior. Like Ward he was greatly influenced by the writings of Comte but, while Comte only talked about the need for empiricism in the study of society, Durkheim became the first notable empirical researcher in sociology. He applied statistical analysis to the study of suicide[6] and his findings on the subject, while open to some question, have also often been supported throughout the past three quarters of a century.[7] For example, Durkheim examined the effect of marital status on suicide rates and found that divorced persons had the highest rates, widowed persons moderate rates, and married persons the lowest rates of suicide.[8] This, among other findings, led him to the conclusion that suicide rates are largely inversely related to the degree of social integration within a society.[9] Utilizing an index of marital

[6]Emile Durkheim, *Le Suicide*, France, 1897. (In English translation see Emile Durkheim, *Suicide: A Study in Sociology*, edited by George Simpson (Glencoe, Ill.: Free Press, 1951.)

[7]See, e.g., Jack P. Gibbs and Walter T. Martin, *Status Integration and Suicide* (Eugene, Oreg.: University of Oregon Press, 1964) and Barclay D. Johnson, "Durkheim's One Cause of Suicide," *American Sociological Review*, XXX (December, 1965), 875–86.

[8]Durkheim (English translation), *op. cit.*, pp. 171–97 and 259–70.

[9]Ibid., p. 209.

integration in a study of suicide in Ceylon, Gibbs and Martin have provided some cross-cultural evidence that Durkheim's conclusion was both useful and reasonable.[10] For Durkheim sociology necessarily had to be the study of social facts. Those data derived from a study of the effects of society on people—the case of group rather than individual action and influence—were seen as the only proper materials for social analysis. In fact, to Durkheim, a person did very little which was not influenced by the various groups of which he was a member. He declared that a prime concern of sociologists must be the determination of the purposes and the consequences of social action. This "functional" approach to social analysis has largely been the major facet of sociological theory ever since and also contributed greatly to the development of theoretical approaches in cultural anthropology.

At about the time Durkheim's works were first being published the University of Chicago was being established in the United States. One of the early members of the Department of Sociology at Chicago was a former language instructor, student of psychology, and journalist by the name of William Isaac Thomas. Thomas became interested in the problem of immigrants to the United States and in conjunction with a philosopher, Florian Znaniecki, he conducted a study of Polish immigrants in the United States.[11] Upon the completion of their study, which introduced the first major use by sociologists of documentary analysis and the life history as data gathering techniques, Thomas and Znaniecki wrote an essay on methodology which they attached to the beginning of their report. Within the context of this "methodological note" (almost 90 pages long) Thomas presented the core of his theoretical contributions to modern sociology. One of those contributions is the concept of "definition of the situation." This concept implies that every person reacts in a social situation according to the way his unique life experiences and, more importantly, his cultural background have oriented him. This idea is the basis of some so-called modern systematic theories in sociology which generally re-

[10]Jack P. Gibbs and Walter T. Martin "Status Integration and Suicide in Ceylon," *American Journal of Sociology*, LXIV (May, 1959), 585–91.

[11]William I. Thomas and Florian Znaniecki, *The Polish Peasant in Europe and America* (Chicago: University of Chicago Press, 1918).

state Thomas' ideas in many more words and considerably less intelligible form. Thomas and Znaniecki also presented a definition of "social disorganization" as a decline in the effect of societal rules upon members of a society. It is also to be noted that Thomas and Znaniecki came to the conclusion that social disorganization and social change are interrelated phenomena.[12] They further marked that both disorganization and change are consequences of shifts in the basic values and attitudes within a society.[13] Znaniecki, a philosopher, was probably largely responsible for their definition of values as any objects—material or nonmaterial of signficance to the members of a social system.[14] On the other hand, Thomas, trained in psychology, was likely responsible for defining attitudes as tendencies to react in a given way to particular values.[15] Both men made later contributions to sociology in general and to specific areas of social investigation in particular[16] but the core of their important contributions is to be found in the "Polish Peasant."

It could be said that modern American sociology dates from the time of Thomas and Znaniecki's fortuitous collaboration. This does not mean that no one else has been of any importance in the development of the field but rather that, following these men, sociology began to come into its own as an important academic discipline in America.

[12]*Ibid.*, see pp. 38–73, *passim.*
[13]*Ibid., loc. cit.*
[14]*Ibid.*
[15]*Ibid.*
[16]For example, see Thomas' brilliant *The Unadjusted Girl* (Boston: Little, Brown and Company, 1923). A classic study of deviant behavior, it illustrates both Thomas' insight as a researcher and his capabilities as a social theorist.

INDEX

Aberle, D. F., 27
accommodation, 31–32
activities, 22, 72–80
acts:
  forbidden, 75, 84–87
  permitted, 75
  required, 75
additives, food, 65
adjusting, 37–42
administration, 78–80
adolescents, 43
aims, 77
alcohol, 84
Alexander, C. Norman, Jr., 91
Allen, Francis R., 106
Allport, Gordon W., 46
Alor, The people of, 45
alternatives, 49
American Anthropological Association, 92
Americans, 44
analysis, social systems, 16–22
Anderson, Alan R., 55
Anderson, Barbara G., 21
Anderson, Robert T., 21
Angrist, Shirley S., 81
anthropology, 3, 38–40, 43–45
apathy, public, 97
apparatus, material, 77
appearance, 42
Applebaum, Richard P., 99
approach:
  applicative, 4, 6, 7
  basic science, 4–7
art, 76, 103
artifacts, 93–97
assimilation, 31, 33
associations, 76
attitudes, 48, 51–53
authority, 65
aviation, 96–97
Axon, Gordon, 70
Aztecs, 84

Babbitt, 69
Bales, Robert F., 25
Bates, Frederick L., 75
Bates, William M., 91
Bateson, Gregory, 12
Becker, Howard, 12, 101
behavior:
  expected, 41–43, 52
  public, 85

behavioral question, 22
Bendix, Reinhardt, 68
Benedict, Ruth, 45
Benoit-Smullyan, Emile, 58
Bentz, Kenneth W., 106
Bergel, Egon E., 62
Berkowitz, Leonard, 47
Bidwell, Charles E., 56
Bierstedt, Robert, 10
birth control, 1
Bittner, Egon, 81
Blacks, American, 68–69
Blau, Peter M., 79
Bock, Philip K., 71
Bodenhafer, Walter B., 1
Boggs, Sara L., 91
Bonner, John T., 12
Borgatta, Edgar F., 16, 25
Boskoff, Alvin, 101
Brand, Donald D., 21
Bredemeier, Harry C., 5–7
Bresnahan, B. S., 25
Briar, Scott, 90
Broyles, J. A., 107
Buettner-Janusch, John, 38
bureaucracy, 77–79
bureaucratization, 77–79
Burgess, Ernest W., 30–31, 50

Caidin, Martin, 96–97
Calhoun, John B., 18
Campbell, Angus, 18
capabilities, inherited, 43
Carneiro, Robert, 82
caste, 62
catastrophe, 89
Census, United States Bureau of,
  59–60, 89
challenge, social, 83–90
change:
  processes of, 102
  progressive, 99
  retrogressive, 99
  social, 92–106
  sources of, natural—biological and
    natural—physical, 93
  technological, 103
Chapple, Eliot D., 14
character, national, 44–45
charter, 76
Child, Irwin L., 45
Christian, John J., 18
Chugarman, Samuel, 110

115

civil rights laws, 88
Clark, Alfred C., 88
Clark, Kenneth B., 107
class (*see* social class)
Cogswell, Betty E., 47
Coleman, A. Lee, 107
Collins, Orvis, 25
communication, 12, 28, 78
community, 20, 22
competition, 31, 33
computers, 4, 100
Comte, Auguste, 108
conflict, 31, 33
conformity, 40–42
constraint, relations of, 43
contact, 12
contravention, 31, 33
control group, 7
Cooley, Charles H., 16, 42–43
Coon, Carleton S., 14
cooperation, 31–32
    relations of, 43
Coser, Lewis A., 8
cultural diffusion, 94–96
cultural proximity, 13
cultural-technological sources of
    change, 93
culture, 37–47, 101
culture levels, 102

D'Antonio, William V., 71
Darwin, Charles, 108
Davis, Keith E., 27
Davis, Kingsley, 67–68
Dean, Dwight G., 25
de Fleur, Melvin, 54
de Gre, Gerard, 76
delinquency, 2, 89
demographic question, 17–19
Dennis, Wayne, 45
density, population, 17–18
deviance, 50, 84–87
    analysis of, 85–87
diffusion, 94–96
Dinitz, Simon, 88
disasters, natural, 93
discoveries, 94
disorganization, 87–89
Drucker, Philip, 37
Dube, S. C., 79
Du Bois, Cora, 45
Dumond, D. E., 106
Dunbar, (Helen) Flanders, 60
duration of interaction, 14
Durkheim, Emile, 18, 52, 87, 111–12
Du Toit, B. M., 107
dyad, 17
dynamics, social, 92–106
Dynes, Russell R., 83, 88
Dynes, Wallace, 35
dysfunction, 29

ecological question, 19–22
ecological relationships, 18–22
economics, 3
economy, 76
education, 75
    level of, 6, 19, 59–60
Edwards, W. E., iii
Eells, Kenneth, 64
efficiency, 78–80
Egypt, 95
Einstein, Albert, 86
Eisenstadt, S. N., 107
elites, 66
Elmer, M. C., ii
Engels, Frederick, 63
environment, total, 12, 93
equilibrium theory, 99–101
Erikson, Kai T., 10, 91
ethnic background, 7, 19, 68–69
ethnocentrism, 110
ethnography, 21
Etzioni, Amitai, 25, 107
Evaluated participation, 64
Evan, William, 82
Ezell, Paul H., 92

family, 71–76
Farber, Maurice L., 47
Fellin, P., 107
Ferguson, Adam, 63
Fichter, Joseph H., 15
Fiellin, Alan, 34
Finer, S. E., 65
Fleisher, Belton M., 91
folkways, 53–54, 110
Food and Drug Administration, 65
Freud, Sigmund, 44
Fritschler, L., 82
frustration, 35
functional prerequisites and
    requisites, 27–28
functions, social, 26–30

Galileo, 86
games, childhood, 41–42
Gamow, George, 86
Gannon, Thomas M., 91
geographic locations, 13, 15, 17, 20–21,
    38–40
geography, 4
Gerth, Hans H., 63, 78
Gibbs, Jack P., 55–56, 111–12
Gillin, J. P., ix, 21
Gittell, Marilyn, 70
Glaser, Barney G., 23
goals and means, illegitimate, 87
goals, long term, 30
    short term, 30
Goffman, Erving, 14
Goffman, Irwin W., 67
Gold, Harry, 6

Goldenweiser, Alexander, 95
Goldfrank, Esther S., 45
Gorer, Geoffrey, 44
Greer, Scott, 78
groups, interacting, 15
  size of, 15–16, 18
Gusfield, Joseph R., 107

Haagensen, C. D., 86
Hagstrom, Warren O., 25
Hall, Richard H., 82
Harper, Dean, 79
Harper, Ernest B., 1
Hassinger, Edward, 23
Hatfield, John, 46
Hausknecht, Murray, 9
Heiss, Jerold, 35
heliocentrists, 95
Henderson, A. M., 63
Henderson, Dan F., 26
Hertzler, Joyce O., 12, 28, 31–32, 75, 78
Heyerdahl, Thor, 95
Hiller, E. T., 59
Hillery, George A., Jr., 24
Himes, Joseph S., 35
history, sociological, 108–13
Hodges, Harold M., 62
Homans, George C., 10
homosexuality, 85
Honigmann, Irma, 45
Honigmann, John, ix, 21, 44
Hooton, Ernest, 38
Horan, Patrick M., 5, 6
Horowitz, I., 10
H.R.A.F. (Human Relations Area Files), 110
Hunter, Floyd, 66
Huntington, Ellsworth, 38–39

ideological change, 103
ideology, 97–98, 103
illegal acts, 86
imagination, 42
income, 64
independent discovery, 94–95
Index of status characteristics, 64
Indians, American, 45, 95–96
individual interpretations, 43
individualism, 21
Infeld, Leopold, 86
Inhelder, Barbel, 42
Inkeles, Alex, 47
inquiry, 1–10
institutionalization, 77–80
institutions, social, 75–80
  facilities of, 77
interaction, 12
  context of, 16
  features of, 16
  forms of, 27–36

frequency of, 14
intensity of, 14
interchangeability of parts, 89
interests, vested, 97
intergroup relations, 22
interpersonal relations, 22
interrelationships, 1
invention, 94, 96–97
Ishino, Iwao, 88

Jackson, Elton F., 67
Jeffrey, C. Ray, 71
Johnson, Barclay D., 111
Johnson, G. B., ii
Johnson, Harry M., 29
judges, panel of, 64–65

Kahl, Joseph A., 68–69
Kardiner, Abram, 44–45
Keller, Suzanne, 66
Kennedy, John F., 98
Kerckhoff, Alan C., 27
key status, 59
Keyfitz, Nathan, 18
kinship, 75
Kluckhohn, Clyde, 38
Knudsen, Dean D., 107
Krauss, Irving, 70–71
Kroeber, A. L., 38
Kunkel, John H., 25

Labovitz, Sanford, 10
lag:
  achievement, 100
  cultural, 100
  organizational, 100
  theories of social change, 100–101
laissez-faire, 21, 109
Landis, Judson R., 71
Larson, Richard F., 56
Lazarsfeld, Paul F., 10
Lear, John, 10
learning, 40–41
Lee, Alfred McClung, 35
Lee, Douglas, 56
Leighton, Alexander, 60
Lenski, Gerhard, 28
Levenson, Beverly L., iii, 40
Levenson, Myron H., 40
Lever, H., 7
Levy, Marion F., Jr., 28
Lewis, Sinclair, 69
life experience, 112
Linton, Ralph, 43–45, 49, 74
Lipset, Seymour M., 68
Lloyd, W. E. B., 86
looking-glass self, 42
Loomis, Charles P., 17, 36
Lopata, Helen Z., 80
Lopreato, Joseph, 71, 107
Lowie, Robert H., 22

Lunt, Paul S., 22
Lynd, Helen M., 22
Lynd, Robert S., 22

MacIver, R. M., 12
Malinowski, B., 39–40, 76–77, 95
Manet, 86
Marquand, John P., 69
Marshall, Tony F., 91
Martin, Walter T., 111–12
Martindale, Don, 28
Marwell, Gerald, 90
Marx, Karl, 63, 108
Marxist ideology, 3
Mason, Alan, 91
Matisse, Henri, 86
Mayo, Selz C., 107
McCullers, John C., 46
McGinnis, Robert, 56
McLeon, Jack M., 47
Mead, George H., 41–43
Mead, Margaret, 44
Meadows, Paul, 81
means, illegitimate, and legitimate, 87
measurement of interaction, 13–14
medicine, 86
Meeker, Marcia, 64
Merton, Robert K., 29, 74
Metraux, Rhoda, 44
Mexico, 95
Meyer, Henry J., 16
Micklin, Michael, 25
military, 29, 76, 96–97
Mills, C. Wright, 63, 66, 78
Minnis, Mhyra S., 22
minorities, 68
mobility:
  horizontal, 61
  social, 61–63
  vertical, 61–63
modal personality, 44–45
Monane, Joseph P., 15, 51
Monet, Claude, 86
Moore, Omar K., 55
Moore, Wilbert E., 67–68, 107
mores, 53–54
Morris, Charles W., 41
Morris, R. B., 91
movements, social:
  conservative, 98–99
  liberal, 98–99
  progressive, 98–99
  retrogressive, 98–99
Mowrer, Ernest R., 82
Murdock, George P., 22

national character, 44–45
Nazis, 3, 87
New Deal, 98
New Frontier, 98
Nixon, Richard M., 98

norms, 52–54, 75
  ideal and real, 52, 88
  prescriptive and proscriptive, 53
Norris, Frank, 69
Nunez, Jose Corona, 21

obligations, 74
occupation, 59
O'Connell, James, 106
Octopus, The, 69
Odum, Howard P., 100
Ogburn, William F. 100
old people, 19, 104–5
Olsen, Marvin, 71
organization, 76
  memberships in, 65
  social, 72–80
overcrowding, 18

Palisi, B. J., 82
Panunzio, Constantine, 75
Pareto, Vilfredo, 65
Park, Robert E., 30–31
Parsons, Talcott, 63, 68, 74
party, 63–64
Patterson, M., 25
Penalosa, Fernando, 25
personality, 43–45
  basic structure of, 44–45
personnel, 77
Peterson, Ronald A., 47
Piaget, Jean, 42–43
Piliran, Irving, 90
Plague, Black, 20
Plant, Walter T., 46
Point of No Return, 69
political behavior, 5–6
political cross-pressure, 5–6
polity, 76
Polo, Marco, 96
Pope, Hallowell, 107
population, 17–19, 102
  change in, 18, 102
  distribution of, 102
Potter, Robert, 25
power, 64–66
power structure, 65–66
prescriptive societies, 101
prestige, 58–60, 63–65
Price, John A., 36
principal societies, 101
privileges, 60
problems, social, 1–2
  sociological, 1
processes, social:
  antagonistic, 32–33
  associative, 32–33
protection, 75
proximity, environmental:
  cultural, 13
  physical, 13

proximity, environmental (*cont.*)
  psycho-physiological, 13
  socio-motivational, 13
psychology, 3
pyramids, 95

Queen, Stuart A., 1
Quinn, James, 22
Quinney, Richard, 90

Ra, 95
Radcliffe-Brown, A. R., 61, 75
Ray, Verne F., 48
recreation, 76
Reed, E. R., 107
Reiss, Ira L., 91
relational networks, 43
religion, 76
required acts, 75
reward and punishment, 53
Robins, Lee W., 71
role, 72–75
  performance, 74–75
  -set, 74–75
  -taking, 42–43
roles, network of, 75
Rose, Arnold M., 51
Rubin, Morton, 57
Ruesch, Jurgen, 13
rules, 77
Ryder, Norman B., 24

St. Simon, Henri, Comte de, 63, 108
Samora, Julian, 71
sanctions, positive and negative:
  economic, 53
  physical, 53
  psychological, 53
Sanders, Irwin T., 22
scale, 51
Scarpitti, Frank R., 6
Schmitt, Raymond L., 82
school systems, analysis of, 79
Schrag, Clarence, 9
Schwartz, Barry, 25
selection of officials, 78
self-feelings, 41–43
Selvin, Hanan C., 25
Semmelweis, Ignaz, 86
sex, 28
Shepherd, Clovis R., 16
Sherif, Muzafer, 11
Shils, Edward A., 74
Simey, T. S., 107
Simmel, Georg, 16–17, 30
Simmons, Leo W., 27
Simpson, George, 110
Simpson, R. L., ix, 68
situs, 58, 73
Slocombe, George E., 86
Slotkin, J. S., 14

Small, Albion W., 17, 30
Smelser, Neil J., 78, 99
Smith, David, 81
Smith, G. Elliott, 95
Smith, H. L., ii
Smith, M. G., 22
social change, 92–106
  situational concerns in, 101
social class, 62–65
  lower. 64–68
  middle, 64–66
  upper, 64–67
social deviancy and social
  disorganization, 2, 83–90
social dynamics, 92–106
social interaction, 2, 11–14, 17–22
social movements, 97–99
social sciences, 3–4
social self, 41–43
social stratification, 61–69
social systems, 1, 14–22, 57–71, 87
social trends, 104–5
social unit, 45
socialization, 40–43
social-psychological sources of
  change, 93
societies:
  pronormless, 101
  proverbial, 101
society, 15, 40, 79
sociological approaches, 4–8
sociology, definition of, 2
  relation to other social sciences, 3–4
socio-psychological reality of
  status, 60
Sorokin, Pitirim, 17
specialties, 49
Spencer, Herbert, 109
Spinden, Herbert J., 95
Spiro, Melford E., 43
standardized norm-sets, 75
status, 57–71
  consistency and inconsistency, 67
  differences, 58–70
  statuses, network of, 60
Stehr, Nico, 67
Stinchcombe, Arthur L., 78
stratification:
  functional and invidious, 66–69
  closed, 61–63
  open, 61–63
  restricted, 61–63
stratification system, feudal, 62
Strauss, Anselm R., 23
Strubing, Carl M., 36
structure, class, 57
Styler, Herman, 20
Sumner, William Graham, 53, 84,
  109–11
superiority-inferiority, 109
Sutker, Sara S., 56

120 *Index*

system characteristics, 16
system levels, 102

Tannenbaum, Percy, H., 47
Taylor, Stanley, 10
technology, 96–97, 103
telic behavior, 29
Thomas, William I., 43, 112–13
Toby, Jackson, 56
Tönnies, Ferdinand, 17
Toynbee, Arnold, 38
traditionalism, 97
Treiman, Donald J., 71
triad, 17
Tumin, Melvin, 68
Turner, Ralph H., 47
Tylor, E. B., 38

Underwood, Francis W., 45
union officials, 19
universals, 49

Vaillant, George C., 84
values, core and social change, 101
values, positive and negative, 50–51
Vance, R. B., ii
Vander Zanden, James W., 27
Vanstone, James W., 72
Vaz, Edmund W., 91

Vexler, Richard B., 103
von Wiese, Leopold, 30
Voss, Harwin, 91

Wallace, Anthony, F. C., 44, 47
Ward, Lester F., 29, 110–11
Warner, W. Lloyd, 22, 64–65
Watson, Jeanne, 25
Weber, Max, 63–64, 78
Wechsler, Henry, 91
Weinstein, Michael A., 9
Weiss, Robert S., 9
Westcott, Roger W., 40
Westie, Frank, 54
Whiting, John W. M., 45
Whyte, William F., 81
Willhelm, Sidney, 25, 55
Williamson, Robert C., 71
Wilson, Mitchell, 94
Wineberg, Morton D., 36
Wolff, Kurt H., 30
Wright, V. C., ix

Yeracaris, Constantine A., 56

Ziegler, Philip, 20
Zimmerman, Carle C., 17
Znaniecki, Florian, 43, 112–13
Zurcher, Louis A., 82